COLUMBIA GLOBAL REPORTS
NEW YORK

Flying Green
On the Frontiers
of New Aviation

Christopher de Bellaigue

Flying Green
On the Frontiers of New Aviation
Copyright © 2023 by Christopher de Bellaigue

Published by Columbia Global Reports
91 Claremont Avenue, Suite 515
New York, NY 10027
globalreports.columbia.edu
facebook.com/columbiaglobalreports
@columbiaGR

Library of Congress Cataloging-in-Publication Data

Names: De Bellaigue, Christopher, 1971- author.
Title: Flying green : on the frontiers of new aviation / Christopher de Bellaigue.
Description: New York, NY : Columbia Global Reports, [2023] | Includes
    bibliographical references.
Identifiers: LCCN 2022046801 (print) | LCCN 2022046802 (ebook) |
    ISBN 9781735913780 (paperback) | ISBN 9781735913797 (ebook)
Subjects: LCSH: Aeronautics--Environmental aspects. | Aeronautics--Technological
    innovations. | Aeronautics, Commercial--Environmental aspects.
Classification: LCC TD195.A27 D43 2023 (print) | LCC TD195.A27 (ebook) |
    DDC 629.130028/6--dc23/eng/20221110
LC record available at https://lccn.loc.gov/2022046801
LC ebook record available at https://lccn.loc.gov/2022046802

Book design by Strick&Williams
Map design by Jeffrey L. Ward
Author photograph by Hugh Gilbert

Printed in the United States of America

For Louisa

CONTENTS

What Is Flying?

Two problems needed solutions before a man could fly. The first involved figuring out the wing curvature—the camber—that would give the craft maximum lift. The ideal camber was identified by the Wright brothers by studying buzzards and testing the glider they had made above their bicycle workshop in Dayton, Ohio. The second concerned power, and was solved by the brothers' mechanic, Charlie Taylor. He built a combustion engine whose main advantage was that it was light. The two elements came together in the Flyer, the biplane that, on December 17, 1903, carried Orville Wright a distance of 120 feet over the sands of Kitty Hawk.

The Wright brothers' first publicist was a bee-enthusiast called Amos Root, of Medina, Ohio. Amos loved his Oldsmobile Runabout, and at $350, it was cheaper than a horse and carriage, as he would say. Hearing reports that the Wrights were experimenting with a flying machine, he drove down to attend trials, and his description of a flight, with Wilbur at the controls, appeared in the trade journal *Gleanings in Bee Culture*. The time would soon come, he predicted, when "we shall not need to fuss with good roads nor railway tracks, bridges, etc., at such an enormous expense. With these machines we can bid adieu to all these things . . . these two brothers have probably not even a faint glimpse of what their discovery is going to bring to the children of men."

The early aviators often returned to themes of destiny and progress, but most of all they spoke of liberation. In the 1920s a young Swiss called Hermann Geiger earned his pilot's license. He didn't dwell on the mechanics of the event, the particular arrangement of wood and cloth and metal that kept him aloft, high over the Alps. The word he reached for to describe his new state of being was "free . . . free to invite people to experience what I had experienced, to take them into the valleys of clouds."

We have learned a lot since then, about freedom, about physics—a century's worth of producing more planes, different planes, flying them farther, faster, and designing them for a wider range of applications, and for many more people. And for most of us who fly today, the question isn't how high, or how far, or how fast, but how much. Down comes the finger and up goes the credit card bill. After a thrill of anticipation that we are going to Rome or Dubai or Tampa, now that we have slammed the door and are heading for the airport, what is it that we are thinking?

The answer is: nothing. When we fly, we volunteer to become a human cargo with a slowed metabolism and depressed expectations. Flying pins us listless and dehydrated to a seat of Lilliputian proportions, unable to move, fed a diet of synthetic entertainment and alleged "food" while being tossed around in a cigar tube at 35,000 feet. And yet, even now, like a dim memory of drawing water from a Neolithic pool, when we look out of the

12 window and see the sun come up over the bowl of the earth and our flying machine is suspended in pure motion, we feel free.

A plane is more than simply a tool of mobility. It's a silver bullet that slays the demon of distance. It's the closest thing we have to supernatural powers, springing open the cage of our immediate existence and propelling us into another. If you're someone who thinks flying's not for you, it's like saying, "I won't leave the cage."

In the early days of aviation, the sky's exploration was both a threat and an opportunity to the nation-state. Sovereignty, hitherto restricted to the contours of the earth, sprang above it, miles above it. The aerospace wasn't something a serious government could afford not to exploit. If you don't, someone else will.

H. G. Wells understood this. He imagined the day when New York City was aflame following aerial bombardment. "The war comes through the air," he wrote. "Bombs drop in the night. Quiet people go out in the morning, and see the air fleets passing overhead—dripping death, dripping death!" Nationalists around the world understood it too. "When the European explorers went to the Arctic and their aviators flew in the skies, we used to laugh at them and say, 'Look how these stupid Europeans get themselves eaten by polar bears and blown to pieces in plane crashes,'" one such nationalist, Cevdet Bey of Istanbul, wrote. "We did not realize that by these 'stupid' acts they realized their domination over the world. Now our men too have begun to crash. This is not something to grieve over. We must rejoice! For me it is the sign that we are regenerating and that we shall not die!"

In the First World War, the belligerents realized Wells's vision of planes in combat. By the end of the conflict, some 200,000 aircraft had been produced, mostly for reconnaissance. But a French company also found that if you mounted a machine gun in a forward position, and put bullet deflectors behind the propeller blades, you could shoot from the cockpit without killing yourself. By 1925, a US federal commission declared that aviation was "vital" to the nation's defense, and that the flying machine should be continuously improved. Subsidies doubled for airmail carriers, the government took responsibility for flyways, and Congress approved a five-year plan for Army and Navy aircraft procurement.

Glamour and convenience vied for priority in the minds of the early passengers, though with the first sign of turbulence, or a thick pea-souper reducing visibility to zero, the delicate equilibrium in the passenger's mind soon gave way to a desperate need for the unmoving land. When Sinclair Lewis's character Dodsworth took to European skies in the early 1920s, at first he was surprised by the bewildering sense of being nowhere, and of nothing happening, when in fact he was—the view from the window confirmed this—perhaps a mile up in the air, and he laughed at his earlier nervousness. His good humor lasted only until his plane ran into a storm, was lashed by rain, and plunged like an express elevator, when two of his neighbors, the first typing at a typewriter, the second sousing himself with Cognac, were sick into paper bags. "Helpless as dice in a box," Dodsworth and his fellow confinees eventually, and "with ineffable gratitude," circled down to the flying-field at Dortmund.

Americans like Dodsworth (and real ones too) were still obliged to sail across the Atlantic to Europe if they wanted to fly,

14 because there were few passenger flights in the US. In the land of its birth, airmail made up most of aviation—until Charles Lindbergh, that is. In the spring of 1927, Lindbergh flew out from Newfoundland with the ambition of crossing the Atlantic by himself, and suddenly the destiny of a single aviator became a matter of vital importance to millions. When the *Spirit of St. Louis* landed at Le Bourget on the evening of May 21, the waiting thousands surged toward the craft, pressing their bodies against its fuselage and tearing off pieces like fragments of a relic.

The Lindbergh effect was instantaneous. Within six months of his crossing, applications for US pilots' licenses tripled and the number of planes quadrupled.

Soon, US airlines were flying more passengers than the rest of the world's airlines put together. Not that the sector was reliably or even generally profitable; the cost of replacing obsolete planes with models that were faster, bigger, safer, and more comfortable made sure of that. The pattern was as follows: the aviation companies invested in order to make more advanced planes, there was a spurt in demand and ticket prices went up, then the airlines bought too many planes, demand slackened, and the loss-makers merged.

From the 1929 crash emerged anti-trust legislation that barred an aviation company from being both a manufacturer and an operator. From then on Boeing, Douglas, McDonnell, and Lockheed would make planes; General Electric and Pratt & Whitney would make engines; and United, TWA, and American would buy their products. The new airliners boasted sleek monocoque fuselages, engines set in the wings to reduce drag, more reliable navigation tools, radio communication, and—miracle of

miracles—pressurized cabins, allowing people to fly above the
weather.

All this happened under the gaze of the state. The early
automotive companies had been named for their owners:
Daimler; Renault; Ford—men with visions and bottom lines.
The early aviation companies took the names of nations, sacred
missions, and pacts between peoples and the elements: Imperial Airways; American Airlines; Pan American Airways; United
Airlines; KLM Royal Dutch Airlines; the Societe anonyme belge
d'Exploitation de la Navigation aerienne; Deutsche Luft Hansa.
An industry that makes the fighters, bombers, transports, and
radar systems upon which the nation's existence depends; that
needs the official permission to introduce retractable landing
gear or to fly from Palermo to Milan; that carries kings and pres-
idents, in all their person and dignity—relations between the
government and such an industry are, by definition, umbilical.

The risk is the technology. Will it work? Will it be safe?
Will it bankrupt the company? You want to grab more market
share, but you don't want to lose what you already have. All this
explains why war—when the government puts up the money
and the red tape drops away—is the best time to improve planes.
During World War II, radar gave Britain advance warning of
aerial attacks on the convoys that were bringing supplies across
the Atlantic. Mass-production techniques allowed the US to
make more and more planes. But the biggest challenge was to fly
faster than the enemy.

The established way to convert energy to thrust was the
Wright brothers' way: burn liquid fuel and use the energy produced to push pistons. But a lot of that energy went to waste,

16 and Newton's third law—that for every action there must be an equal and opposite reaction—suggested a more efficient alternative. In the 1930s, a British engineer and aviator named Frank Whittle compressed fuel into a turbine and produced a powerful jet. At first, the Royal Air Force was put off of developing Whittle's idea by the cost, but war stiffens resolve. By 1945, not one but two jet fighters, Britain's Gloster Meteor and Germany's Messerschmitt Me 262, were streaking through European skies at unprecedented speeds. By the 1950s, the jet engine was all but mandatory for combat aircraft.

Civilian aviation is also about speed, though the goal is not to save lives but to save time. The flying public regarded crossing the Atlantic in a propeller-driven Douglas DC-3 with a refueling stop at Iceland as a less-than-ideal way to spend twenty-six hours. In 1952, the British manufacturer de Havilland launched the world's first jet airliner, the Comet. The Comet was two and a half times faster than the DC-3, but it was unsafe.

On October 26, 1958, while the Comet was grounded following its umpteenth crash, a bigger, faster, safer jet plane, the Boeing 707, took off from New York. After receiving 17,000 gallons of fuel in Newfoundland, the 707 took just seven hours to reach Le Bourget. Its 111 passengers—the largest complement ever aboard a regular passenger service—were in raptures over the smooth, quiet conditions in a cabin that was, in the words of one reporter, "almost as wide as a living room and as long as a ballroom."

Over the next forty years, Boeing and a younger European competitor, Airbus, formed the duopoly that now accounts for 99 percent of the world's large aircraft orders. But the expertise, money, and political power that was concentrated in the two

giants didn't translate into a welter of design improvements.
When it came to transformative technologies, the jet engine
would be the last.

Modern planes are built with millions of parts sourced from
thousands of suppliers and made from a huge variety of
materials. Every design change, however small, needs regulatory approval. "Clean-sheet" design, which means developing a new aircraft from scratch, is phenomenally expensive,
time-consuming, and risky.

Studies for the world's largest passenger plane, the Airbus
A380, started in 1988. By the time it entered service, in 2007,
behind schedule and over budget, its technology was already
out of date and its guiding premise, that bigger was better, had
been discredited. The A380 was too big to fill and too expensive to buy—its $445 million price tag wasn't sufficient to
cover production cost, let alone recoup the estimated $25 bil-
lion Airbus had spent developing it. The program was finally
put out of its misery in 2021, its debts to European governments unpaid.

For airlines nowadays, the best new plane isn't a new plane
at all. It's a more efficient iteration of a plane they already know,
and which they don't need to retrain pilots and technicians in
order to use. For decades, innovation has been directed pri-
marily at cutting operating costs. Winglets increase aerodyna-
mism and reduce flight times. Computers obviate the need to
fill the flight deck with expensively trained, well-paid human
beings. Intelligent scheduling also helps. (Have you noticed
how holding patterns aren't the thing they used to be?) By tin-
kering with engines, deadweight, drag, even cutting the weight

of the in-flight magazine, the airlines have been able to improve fuel efficiency by at least 1 percent a year.

Tight margins—the bane of the modern aviation industry—are a consequence of liberalization. Until the late 1970s, fares and routes were determined by governments. The airlines, many of them national carriers enjoying monopoly status, were content to keep prices high and serve only a tiny minority that could afford the ticket price. Then—first in America, then in the EU, and later in growth-markets like India—aviation was deregulated and became viciously competitive. New low-cost carriers like Southwest and Ryanair packed more and more passengers into their planes and swapped frills like food and legroom for a lower ticket price. For travelers prepared to fast, carry only hand baggage, and squash into a middle seat, the cost of flying dropped dramatically. In 1960, a one-way flight between New York and London would have cost you around $300. If you shop around now, you can travel the same route for the same price, despite the fact that inflation has depreciated your $300 by more than 900 percent.

Squeezed on the one hand by the public's insatiable appetite for budget travel, on the other by a long-term (if uneven) rise in fuel prices, the airlines borrowed accordingly. Between 1981 and 2000, the airline industry was in the red every single year. The industry survived by doing what every low-margin, heavily leveraged sector of the economy does. It scaled.

If you look at the graphs that the International Air Transport Association (IATA), the airlines' trade body, puts out to demonstrate how indispensable flying has become, the thing that stands out is that aviation is about numbers. The line showing the number of passenger journeys taken per year is all but asleep in the 1950s and 1960s, wakes with a jolt in the 1990s, and positively soars after the new millennium. This being an industry that is perhaps uniquely susceptible to external disruption, every now and then the line flattens. These plateaus correspond to the 1979 oil shock, the first Gulf War, 9/11, and the 2008 financial crisis. But they are strictly temporary— pauses for breath while the industry retrenches, borrows (from lenders), begs (from government), and dusts itself off for the next near-as-dammit-vertical take-off.

Since 1970, the underlying growth of air traffic has averaged 4.4 percent a year. The dream year was 2019, when a total of 4.5 billion passengers were flown (a 3.6 percent rise on the 2018 figure, itself a 6.4 percent rise on 2017) and 82 percent of seats were full. The biggest increase came in the Asia-Pacific region (nearly 10 percent), which also accounted for the biggest share of passenger flights, 35 percent. The supposedly "mature" market of Europe (a quarter of world traffic) posted an increase of more than 7 percent.

Persuading people to fly more often isn't the same as persuading more people to fly, though it suits the industry to blur the two. According to IATA, in 2017, "worldwide annual air passenger numbers exceeded four billion for the first time," and "the average citizen flew . . . once every twenty-two months." But this doesn't mean that more than half the world's population flew that year; the number refers to the total number of flights, including return and connecting ones, and taken by a much smaller number of people. Nor are there many "average citizens" out there, flying stolidly but infrequently. Rather, there is a tiny number of rich people who fly very often, a bigger number who fly several times a year, and a great many who go for

20 years without flying at all. Researchers have estimated that the 823 million international flights recorded in 2018 were taken by a mere 155 million people—just 2 percent of the world's population. Impressive-sounding "passenger numbers" allow the industry to propagate the notion that air travel is a "norm" that cannot be dispensed with, involving a larger proportion of the global population than is actually the case.

Critics of the volumes-based approach to flying have often predicted that aviation will one day be discovered under the wreckage of its own business model. There was even a short-lived theory that flying would be rendered obsolete by the internet's ability to replicate lived experience. In his book *The End of Airports*, Christopher Schaberg suggested that "in a world where . . . connections happen as easily online as off, it seems inevitable that moving hundreds of bodies around in large vessels will go out of fashion."

If there was ever a time for Schaberg's thesis to come true, the COVID pandemic was it. In May 2020, passenger demand was down 94 percent on May 2019. The industry was on its knees. But aviation didn't die. As I write these words, in the summer of 2022, it is coming rapidly to life, even in the face of airline staff shortages and airport chaos, the war in Ukraine, and the rocketing price of fuel. According to IATA predictions, traveler numbers will rise to 83 percent of 2019 levels in 2022, 94 percent in 2023, 103 percent in 2024, and 111 percent in 2025. Whether or not this optimism is justified, a return to stratospheric volumes-growth is indeed essential if the industry is to outpace its creditors.

If the global aviation sector were a country, its total contribution (direct, indirect, induced, and catalytic) of $2.7 trillion to the gross domestic product, and the 65.5 million jobs it supports, would be comparable to the United Kingdom's economic size and population. These figures were arrived at by the International Civil Aviation Organization, the UN body that sets the rules for international aviation. And without aviation, the exponentially larger tourism sector, which in 2019 accounted for one-tenth of global GDP and employed 334 million people, would effectively cease to exist.

Around one in 130 American workers is employed either by Boeing, with its domestic payroll of 143,000, or by one of its 12,000 local suppliers, with another 1 million workers. Airbus has 10,000 to 20,000 fewer employees, but a supply network of similar size and complexity. The largesse that the US government extends to Boeing, and European governments to Airbus, is that of parents terrified of upsetting their young. Perhaps unsurprisingly, given the sector's clout, when the pandemic struck in 2020, some $243 billion was conjured in short order from public and private sources to rescue the airline industry.

In 2019, the Industry High Level Group, an umbrella body that brings together airlines, airports, manufacturers, and other players, noted that "despite the long-lasting and vital importance of air transport to global development, the expansion of aviation today faces many challenges and indeed threats. Continued political support and economic investment will be needed in the aviation sector to meet its potential." Two years later, as the industry limped out of the pandemic, Willie Walsh, IATA's director general and a former CEO of British Airways, pointed out that "the magnitude of the COVID-19 crisis for airlines is enormous. Over the 2020–2022 period total losses could top $200 billion . . . the scale of this crisis needs solutions

22 that only governments can provide." The industry's approach boils down to this: give us the money and we'll give you the volumes. For it's volumes—not innovation; not the promise of a sustainable business model; not, heaven forbid, blue sky thinking of any kind—that constitute aviation's "potential."

But then a body of evidence arrives that shows that the aviation sector isn't simply an economic and social oddity, but is in fact exceptionally prejudicial to the health of the world. Once it's understood that aviation's contribution to climate change is grossly disproportionate to the number of people who use it, and that this contribution is set to increase dramatically, then it becomes a problem not just for the 11 percent of the world's population who fly in a single year, or the 1 percent who account for more than half the total emissions from passenger aviation (and are handsomely rewarded by the airlines for their "loyalty")—but for everyone, particularly those who are condemned to live in the sinking or desiccating or monsoon-failing fringes of the non-airborne world, the so-called "front-line" of climate change.

Unlike ground vehicles, aircraft don't have to worry about rolling resistance. Modern planes look the way they do, with snub noses and sleek bodies, because aircraft designers have learned to minimize drag. The result is that an airliner with a full complement of passengers has a fuel efficiency of 30 to 35 kilometers per liter of fuel per passenger. That's half the efficiency of a car, which isn't bad when you consider that a plane's job description includes defying gravity.

But comparing a car and a plane is like comparing a gnat and an elephant. They range and excrete on different scales. A long car journey of, say, eleven hours, with breaks for fuel and other necessities, might cover 650 miles. In a full car, the amount of fuel burned by each passenger is under 14 kilograms. In the same eleven hours in an airliner, you've gone from Paris to San Francisco, a distance of 5,500 miles, and the average amount of fuel burned by each passenger exceeds 300 kilograms. And that's assuming the plane is full. If it's half-empty, the per-person burn is much higher. If you're in business class, the figure is higher still. In first class, it's off the scale.

The fuel that's being burned is known as jet fuel. You and I call it kerosene. Burning kerosene releases greenhouse gases and other drivers of global warming. The amount of fuel that is burned in a plane engine—thus, the amount of global warming it precipitates—depends on factors that include the efficiency of the motor, the aerodynamics of the craft, how heavy its load is, the altitude at which it is flying, and the distance it travels. Take-off is the phase of flight that needs the most energy. Cruising needs the least. But that doesn't mean that the longer the cruise, the more efficient the energy use. Like camels in the desert, aircraft carry their fuel around with them; a long journey needs a lot of kerosene, and kerosene is heavy. A lot of the fuel in the tank is there simply to carry the fuel in the tank.

In 2018, CO_2 emissions from civil aviation represented around 2.4 percent of total global emissions caused by human activity. But the impact of aviation on the climate goes beyond CO_2. In fact, CO_2 emissions may constitute little more than a third of the sector's contribution to climate change. Aircraft also release nitrogen oxides, oxidized sulfur, water vapor, and contrail cirrus—artificial clouds caused when water vapor condenses around soot from the plane's exhaust at high altitude.

24 This temporarily increases the amount of heat that is trapped. All of these affect the climate; their combined effect is to warm it—in the case of contrail cirrus, much more intensively, if more briefly, than CO_2. Released at high altitudes, aviation emissions have between two and four times the impact of comparable ground source emissions.

If we take these additional impacts into account, we see that aviation represents around 3.5 percent of the warming impact caused by humans. That compares to around 6 percent for the cement sector and 17 percent for cars. But cement is in virtually every road and building that's made, and cars are vital to billions. For all its pursuit of volumes, aviation barely counts as a mass activity, and rarely as an essential one. The majority of people who take planes do so not for vital work or family reasons but in order to have fun at the other end. Aviation has a strong claim to be the most damaging leisure activity around.

When you fly, you emit up to 100 times more CO_2-equivalent per hour than you do if you travel by train, bus, or in a fully occupied car. Flying from Berlin to Vancouver and back again, you emit in a few hours more greenhouse gases than the average non-airborne Indian or Nigerian—i.e., the majority of the inhabitants of two of the world's most populous countries—does in a year. In fact, air travel is almost certainly your biggest single contribution to greenhouse gas emissions, dwarfing your domestic energy use, daily commute, hobbies, and diet. In the words of Andrew Murphy, an aviation expert at Transport and Environment, a Brussels-based NGO, "euro for euro, hour for hour, flying is the quickest and cheapest way to warm the planet."

As a global sector, transport lags behind global climate targets. Within transport, aviation is even further behind, sharing a subset (with shipping) of spectacular underachievers. To return to the comparison with cement and cars, emissions from both of these sectors in Europe and the US are on a sharp downward trajectory, while emissions from aviation are expected to continue rising sharply after the hiatus caused by COVID.

A recent study by researchers at the Universities of British Columbia and Lund identified giving up one transatlantic flight per year as one of four actions we could take that would have the greatest positive impact on the environment. Compared to the other three—having one fewer child, eating a plant-based diet, and giving up travel by car—it would also, for most, be the most straightforward. But governments rarely tell people the truth about climate change. They don't want to be the bringers of bad news. The same study found that science textbooks in Canadian schools barely mention these high-impact actions when dealing with climate change, focusing instead on lifestyle choices that have hardly any impact at all. One textbook claimed that "making a difference doesn't have to be difficult" and used the example of switching from plastic bags to reusable shopping bags in order to save five kilograms of CO_2 per year. It didn't mention that this would achieve less than half a percent of the savings that result from choosing not to fly once across the Atlantic and back.

Pandemic aside, I cannot think of a year since I was seventeen years of age when I haven't flown, often ten times or more. Working as a foreign correspondent and living abroad, desiring the pleasure of a holiday or to see people I am close to, it seemed like the obvious thing to do.

26   The strongest emotion it aroused in me was fear. A glass of champagne and a seat designed for a fully formed human being were the only acceptable antidotes to the fright and rebellion that tightened around my chest at take-off, when I looked around at the placid faces and saw that I was the only person who realized how unwise we were to divorce ourselves from the good earth and that we were all going to die. To put it another way, my fear of falling out of the sky was sharpened by the extremes of deprivation to be found in economy.

Climate change mugs us in phases. First it's disbelief at the initial data, then unease as the data ripens into a thick body of evidence (it would certainly be inconvenient if this turns out to be true), and on to frustration at the inaction of politicians, false dawns, defiance (*carpe diem*), resolve (cut up the airline credit card but not before using up the points), impotent railing (we, God's elect, Darwin's winners, soiling our own nest!), culpability, and . . . a deeper, more implacable unease.

Have you ever found yourself whiling away a long flight by watching a documentary in which David Attenborough reads the last rites over planet Earth? You too? Our every private irony, our every secret pang, is in fact embarrassingly banal.

If you want a good, succinct account of the personal dilemmas and choices that anyone who is considering the question of aviation inevitably faces, I recommend looking at a Ted Talk called "We Need to Stay on the Ground," by a woman named Maja Rosen. In this talk, immediately datable to peak-COVID by the fact that she isn't striding around a stage somewhere, but is seated, alone, against a dark backdrop, a woman with buoyant hair, fine blue eyes, and the better-than-native English that

Scandinavians imbibe with their mothers' milk, slowly and reasonably lays out the interest that we all have in cutting down on our flying. She examines the convenient ignorance that people exhibit when it comes to knowing about the damage that flying does, flying's elite status ("The reason that aviation emissions aren't higher is because very few people in the world are privileged enough to fly"), and the atonality of public messaging on the subject ("In the media, news about the climate sits next to articles about cheap holiday flights"). But Rosen doesn't lecture, much less sneer. She's keen not to alienate anyone she might ultimately convince. "The reason so many of us continue to fly away on holiday," she says reassuringly, "is not because we don't care, but because we are human."

If being human means we are drawn to do what others do, it follows that if others choose not to do something, we might feel drawn to emulate that too. "Choosing not to fly . . . is a very effective way to get more people to wake up and realize the severity of the situation," is how Rosen puts it. "And the same applies to the reverse scenario. If you continue to fly, you signal to people around you that everything is under control and that they can calmly continue to ignore that we're in a climate emergency. Giving up flying is one of the most important things that you can do as an individual to reduce your own emissions, but the biggest effect lies in how this decision impacts others."

A Swede in her early forties, Rosen used to live and work in the UK, I found out when we spoke by Zoom a little later in the pandemic. Ryanair kept her in touch with home and allowed her friends to visit. A classic case of travel—in this case, the displacement of one person across the North Sea—begetting more

28   travel. Then, in 2007, came a trip by plane to the Arctic Circle, melting before her eyes, and Rosen resolved to fly no more.

The stages she went through—ambition, hopelessness, intelligent resolve—will be familiar to many climate worriers. "At first I was terrified," she remembered. "And I tried straight away to convince all my friends to stop flying too. But I couldn't talk about this in a constructive way. I realized that they weren't listening and I just got frustrated." Then came the slump. "I kind of gave up, I felt like there was no point. It was after I had my second child, and I was on parental leave, and I remember a few times when I was talking to other people who had recently become parents. And we were sitting there with our newborn babies, and they would say, 'Oh, you know, I can't wait to go to Spain' or wherever. And I went home, and I couldn't sleep because I felt like such a coward, because I was more scared of being a killjoy than of the climate collapsing and humanity being destroyed." Finally, in 2015, Rosen took a new year's resolution "to be brave and ask questions."

Rosen's opportunity to apply her resolution came one day when her landlord knocked at her door and asked if she would look after his cat while he and his wife went on holiday to Vietnam. "And although part of me was thinking, 'He's my landlord and it's really important for us to have a good relationship,' what I ended up saying was, 'Aren't you worried about the climate crisis? And are you aware of how big your emissions will be from this trip?'" Having heard her out politely, Rosen's landlord went to Vietnam anyway. A few months later there was another knock at the door. This time it was the landlord's wife, who told Rosen, "We've been thinking over what you said and you were right." Husband and wife both signed up to Rosen's campaign.

Embarrassment—a fear of being looked at with eyes that say, "And who do you think you are to tell me how to live my life?"—is a key accomplice of climate change. And Rosen was ahead of the curve. Not until 2018 did her compatriot Greta Thunberg bring the climate emergency to worldwide prominence with school strikes that were emulated around the world; Extinction Rebellion was set up the same year. By that time Rosen had refined her techniques, and the NGO she had started, We Stay on the Ground, was convincing more and more Swedes to do precisely that. Not by lecturing them, but by asking questions.

Shortly before the pandemic, even as air travel generally boomed, there were signs that flight shame was stirring some consumers to rebel. Between January and October of 2019, Sweden recorded an average decline in air travel of 8.7 percent compared to the same period in 2018. More significant still, in view of its elevated passenger numbers, air traffic in neighboring Germany grew more and more sluggishly as the year progressed, before going negative in September. In November 2019, passenger levels in Germany showed a decline of 3.7 percent against the same month in 2018, while domestic air travel fell by 12.3 percent. "Could you take the train instead?" KLM asked its passengers in an ad. "We all have to fly every now and then. But next time, think about flying responsibly."

The environmental crisis has made "polluter pays" a fact of life for many economic sectors. Whether it is through carbon pricing for power plants, fines for farmers who pollute rivers, or visitor taxes to mitigate the effects of overtourism, the principle that the cost of environmental degradation must be borne

30   by those who create it has been widely, if grudgingly, accepted. Not, however, by the aviation industry. Not even by KLM, whose Paris-based parent company, at the time of the "fly responsibly" ad, was vigorously criticizing the French government's introduction of a 1.50 euro carbon ticket tax. The customer whom aviation treats the best, whom the airlines garland with its most egregious toadying, the frequent flier, is the worst polluter of all. And the most logical way of reducing aviation's contribution to global warming, which is for people to fly less, is the one that the airlines—and their political allies—steadfastly refuse to entertain.

The aviation industry revels in its own exceptionalism, which is indivisible from its sense of entitlement and is reflected in the bizarre privileges it enjoys. The oldest of these dates back to 1944, when the Allies, hoping to stimulate commercial aviation after the war, exempted it from fuel tax. To this day aviation has no equivalent of the tax that, when you drive a car in the UK, for example, accounts for more than a third of the price you pay at the gas pump. Nor is sales tax levied on international air tickets. Like shipping—another awkwardly mobile sector that scorns national jurisdictions—cross-border aviation is absent from the 2015 Paris Agreement on climate change, in part due to the difficulty of assigning responsibility for the emissions of international flights in which a carrier from one country flies from a second country to a third.

In 2016, the International Civil Aviation Organization (ICAO) agreed to the outline of a Carbon Offsetting and Reduction Scheme for International Aviation, or CORSIA, which aimed to stabilize emissions at 2020 levels. Under CORSIA, airlines buy offsets—planting trees, investing in solar farms, or distributing low-emission stoves in order to atone for any excess growth in carbon emissions above a baseline. CORSIA is hated by environmental groups, which estimate that a lot of the emission reductions that take place under the scheme—for example, avoided deforestation—would have happened anyway. Even after it becomes mandatory, in 2027—at present, India, China, and Russia are among the countries that haven't joined—it won't include domestic flights, which produce more than a third of the industry's emissions. Nor will it take into account non-carbon impacts like contrails. Extrapolating from CORSIA's baseline (which has already shifted in the airlines' favor), analysts predict that over the scheme's anticipated lifetime, between 2021 and 2035, airlines will offset at most 21.6 percent of their emissions.

It has become steadily harder for aviation to remain aloof from the global campaign against climate change. In 2020 some EU countries actually made rescue funds for their COVID-hit carriers dependent on emissions-savings, while the UK became the first major economy to include aviation emissions in its national carbon budget. The following year, IATA passed a resolution "to achieve net-zero carbon emissions by 2050. This commitment will align with the Paris Agreement goal for global warming not to exceed 1.5°C." Finally, in October 2022, the members of ICAO, which sets standards and recommended practices for the industry, adopted a "collective long-term aspirational goal (LTAG) of net-zero carbon emissions by 2050."

In the words of Willie Walsh, "The world's airlines have taken a momentous decision . . . with the collective efforts of the entire value chain and supportive government policies, aviation will achieve net-zero emissions by 2050." But before we get

32 carried away by IATA's resolution and ICAO's non-binding goal, it's worth recalling that both objectives give the industry license to carry on increasing emissions until CORSIA expires in 2035. The idea is that during this period airlines, manufacturers, fuel suppliers, and governments will work together to bring forward technologies that will enable rapid emission reductions after that date. But in light of aviation's volumes-based model and its glaring failure to invest in abatement technologies that in other sectors have been in development for decades, environmentalists' response to the industry's new targets has been along the lines of, "We'll believe it when we see it." Even a supporter of aviation like Richard Aboulafia of AeroDynamic Advisory, an aviation consulting firm, doubts the sector's capacity to leverage a plethora of unproven technologies in very short order. "Barring the help of space aliens," he told me when we spoke in the summer of 2022, "we're going to be the very last industry on the planet that decarbonizes."

Within the hard-to-abate segment, it's possible to argue that aviation is in a class of its own. Even the optimists don't know if decarbonization is possible, let alone which of the available technologies stand the greatest chance of success. Among hydrogen's advocates, there's a division between supporters of the fuel cell and combustion. Proponents of a kind of synthetic fuel called e-fuel consider biofuels to be a positive menace. All unite to deride electrification—all except those investors who have poured billions into electric flight. Some see a place for carbon capture and sequestration. Others regard it as an expensive irrelevance. Everyone ends up saying the same thing: *My method is going to win.* But—assuming that no outside factor intervenes to reduce passenger numbers drastically—none of these on its own will be able to carry the massive burden that has been wished away for so long. If aviation is to be saved from itself, they must all win.

And here is the silver lining. Thousands of startups and new units within companies are pouring resources into technologies that have made flying green their objective. There's a chance that the cumbersome, needy, petulant, change-averse behemoth that is modern aviation is starting to rediscover the fearlessness and zest of the Wright brothers and Whittle, and that in saving itself it will help save the world. Close your eyes, imagine you're back with Charlie Taylor at his workbench in Dayton, and you're tapping him on the shoulder: "Charlie, this engine of yours is all very well, but what other ways exist to get a machine to fly?"

Fuel

The reason hydrocarbons make such good fuels is that a combination of dead plankton, heat, and time creates a lot of energy. Jet fuel has a very high energy density—much higher than lithium batteries, for instance, which is why electric aviation is such a long shot. Making jet fuel—drilling for oil, then processing and refining it—itself releases a lot of greenhouse gases, meaning you're a substantial emitter even before you pour it into the plane. When you burn it, you emit much more. So you're a double sinner. On the other hand, if you draw your hydrocarbons not from underground but from the air, and they end up back in the air, you're neutral when it comes to carbon and sulfur, but still a sinner with respect to contrails.

The city of Zurich is where the Protestant reformer Ulrich Zwingli launched his offensive against Catholicism and its moral offsets in the sixteenth century, a campaign that is remembered arrestingly in a stone relief on a church wall showing a young man with a pudding bowl haircut preaching before a congregation in varying states of puzzlement, ecstasy, and perturbation.

A city with such a history lends itself to gradations of virtue.
And now, in low-slung, residually industrial premises high
above Lake Zurich, a place that hums with its young workforce
moving between research labs, office space, and cavernous ele-
vators, I am learning from an impatient modern reformer called
Christoph Gebald how to make fuel out of thin air.

It's the end of 2021, and the pandemic is becoming merely a
nuisance. Europe is opening up, capriciously and bad-temperedly,
and rather than risk being turned back by the French authorities at the Eurostar terminus in Paris, I have taken to the skies
again, which has left me, if not quite penitent, then certainly
rueful in an unsuccessfully-reforming-alcoholic sort of way.
But Gebald isn't the kind of environmentalist to give me a hard
time. "The global exchange of cultures is something fundamentally important to keeping humanity alive," declares this slim,
soft-spoken man whose fitted pullover and neat black hair give
him the air of a precocious schoolboy. "Therefore, I declare aviation to be something unavoidable; you have to fly." This makes
me feel better.

In 2009, Gebald and his fellow-German Jan Wurzbacher
started a company called Climeworks. The company has since
become famous for pulling CO_2 out of the air in Iceland for injection deep into the volcanic basalt, a gold-plated offset scheme
that numbers Microsoft and the Economist Group among its
clients. But Climeworks started life as a fuels company and its
core technology is essential to the production of sustainable
aviation fuels, otherwise known as SAF.

The principle of pulling air apart and using its constit-
uent gases to produce something else has been understood ever
since the early twentieth century, when German chemists first

36 extracted nitrogen from the air and combined it with hydrogen to produce ammonia fertilizer. "It's really quite low-tech," Gebald tells me with the easy self-assurance that seems to be his hallmark. "You're mimicking nature, right? You close the carbon circle. You take CO_2 out of the air, you make a fuel, and when the fuel is burned, the CO_2 gets re-emitted into the atmosphere. That's one thing. And the other thing is that because the fuel is synthetic, it burns a lot cleaner. So you have almost no pollutants."

Gebald's job is to capture the CO_2. "Essentially we have designed an acid-alkaline reaction, where we immobilize CO_2 on the surface of a filter. Then we reverse this reaction by introducing heat, and pull the CO_2 out of the system with a pump, obtaining a gas so pure you can put it in that drinking bottle in front of you. Or you can use it for the production of renewable fuels."

It all sounds remarkably simple, but the hard part is to make the process energy- and cost-efficient. "Society and decisionmakers don't understand that, even today, if in twenty swim lanes where we have work to do, we went all-out, it's still not clear whether we would make it and overcome the climate crisis or not." He shakes his head dismissively. "And we're still discussing . . ."

What's stopping sustainable aviation fuels from scaling as fast as the severity of the crisis demands? "Why, after twenty years of commercial development, do we still only supply 4 percent of global energy through renewables? Why, in twenty years of solar energy, did we scale capacity only by a multiple of one thousand? Why not ten thousand? There's a law governing how fast you can scale technology. It has to do with execution, funding, and your supply chain."

It also has to do with governments and their willingness to incentivize companies that invest in expensive new technologies and/or penalize those that persist with old, cheap ones. In 2021, under the "Fit for 55" climate package, the European Commission proposed minimum SAF blending volumes in aviation fuel, rising from 2 percent in 2025 to 5 percent in 2030 and 63 percent in 2050. In the United States, the Sustainable Aviation Fuel Grand Challenge aims to scale up SAF production to 11 billion liters a year by 2030, and to meet the country's entire aviation fuel demand by 2050. SAF is a "drop-in" that can be blended with fossil-jet fuel and poured into the tank today. There's no need for clean-sheet technology or new infrastructure at airports. It's little wonder that the industry loves it. Fuel producers like Shell, airports like Amsterdam's Schiphol, and airlines everywhere have all committed themselves to (respectively) producing, retailing, and buying increasing volumes of SAF.

All that stops it from immediately taking a big share of the aviation fuels market is lack of supply. According to the World Economic Forum, in 2019, the industry's high-water mark, fewer than 220,000 tons of SAF were produced globally, representing less than 0.1 percent of the estimated 330 million tons of jet fuel that were used by commercial airlines that year. Even if all SAF projects that have been announced materialize as planned, capacity will still fall short of the 2 percent share that the International Air Transport Industry has targeted for 2025. Ed Bastian, Delta's CEO, observed that if they only used SAF, in a single day his airline would use up all that's produced in the US in a year.

The walls of the corridors at Climeworks are adorned with photos of prototypes and pilot plants—milestones in

38 Climeworks's ascent to a point where it can attract half a billion dollars in a single funding round. "You're a man in a hurry," I suggest. "You're not one of those people who says by running around being human beings we've created all these problems and the best thing to do is to stop running. You're saying the best thing to do is to run faster."

Gebald allows himself a rare pause for breath. "I have several times been at a real low in my life through health and accidents," he says slowly. "And once you're in really deep shit, you understand that, well, I'm a human being, I'm lucky enough to be on this planet for eighty years. And I want to get something out of that. And you have to provide solutions to balance a certain life quality with making sure we're not driving ourselves to extinction. I would love to eat no meat, and I eat very little meat, believe me. And I fly as little as possible. But I don't know whether we will win over 100 percent of the population. I am convinced that technology can do a lot. You know, we're building 100 million cars every year at a cost of less than $10 per kilogram of weight. I find that very, very inspiring."

After I say goodbye to Gebald, I drive an hour or so outside Zurich to Hinwil, a nondescript village where Climeworks has attached a facility to a plant that burns municipal waste. The first principle of sustainability is that nothing is thrown away, and, rather than be released into the atmosphere, the surplus heat that the waste plant generates is directed to a bank of units, each one resembling a large, old-fashioned tumble-dryer, that Climeworks has put on the roof of the plant.

The tumble-dryers are in fact direct air capture machines and what resembles the door where you can see your clothes going round and round is the duct that pulls air into the unit

and over the filter that absorbs the CO_2. Every so often heat is $\quad$
blasted at the filter to release the captured CO_2, some of which is
siphoned into tanks to provide fizz for a local mineral water and
some piped a few hundred yards away to boost yields in an enormous greenhouse.

What's happening on the roof of the municipal waste plant
is a subset of direct air capture called carbon capture and utilization. It is also the first step toward making synthetic fuels
from thin air, or e-fuels. Step two will be to combine the captured CO_2 with water and bring it into an electrolyzer, which
splits the molecules and makes syngas, a gaseous mixture of
hydrogen and carbon monoxide. Finally, using a reaction called
the Fischer-Tropsch synthesis, the carbon and hydrogen will
be fused to produce the same carbon chains that, when refined,
make up the fossil fuels we use to power airplanes.

My guide to Hinwil is Climeworks's head of climate policy,
Christoph Beuttler. He used to work for an environmental think
tank before joining Climeworks—to make things happen, he
tells me. He says that e-fuels will emit around 85 percent fewer
greenhouse gases than today's jet fuel. "And then you can do an
additional removal to get it to zero emissions. The top-up could
happen in Iceland or anywhere else where carbon capture and
sequestration have been rolled out. We could remove even more
than is needed," Beuttler enthuses. "We could even make it climate positive." When we come into the car park before going
our separate ways, Beuttler winces while unlocking a huge,
less-than-climate-friendly SUV. Raising his hands, he says,
"Not my choice," in a way that suggests a family negotiation has
taken place and he has come out the loser. I smile sympatheti-
cally. We all know about compromise.

40   Climeworks is part of a Norwegian-based consortium, Norsk e-Fuel, that will bring together the technologies that go to make e-fuel. The lead partner in this joint venture, the pivot around which the whole proposition revolves, is the German company Sunfire, a leader in electrolyzer technology.

Electrolysis is a means of splitting water into hydrogen and oxygen using an electric current. It is the opposite of a fuel cell, which passes hydrogen through an electric circuit to make water and heat. Since pure water doesn't conduct electricity well, it needs another chemical, called an electrolyte, that gets the process going; alkaline electrolyzers use an alkalite electrolyte. Sunfire has expertise in alkaline electrolyzers, but is also developing a new kind of electrolyzer, called solid oxide, that processes steam and CO_2 into syngas in a single step at very high temperatures.

Sunfire is headquartered in Dresden, a city with a unique insight into the destructive capacity of fossil fuels. Over a few days at the end of the Second World War, British and American bombers dropped 4,400 tons of high explosives onto Dresden's Baroque city center, the very stones of its churches and palaces melting in the inferno, their iron fixings deliquescing into pools. After the war, over several decades, the city center was methodically put back together, a process of restoration with few parallels in human history.

And now, a short distance from the Dresden Panometer, a vast cylinder that was built in the 1870s to store the city's gas reserves, and was recently converted into an art gallery, the atmosphere around the conference table at Sunfire is charged with ideological static. "Oh yes," Carl Berninghausen, Sunfire's chairman and co-founder, assures me, "of course we all want to have breakfast, we all want to have a bed which is warm and dry. So some sort of economic success is absolutely on our agenda. But we don't want to make a profit at the expense of the environment or of people. We want to do it in the proper way. We have a very high birth rate among employees at Sunfire. And as soon as you have children, you think about the future a bit differently."

Here is a feature that is common to many of the entrepreneurs I met while researching this book: their insistence that, while making money is far from disagreeable to them, their primary motivation is moral and environmental. In Zurich, Christoph Gebald told me that Climeworks is "absolutely not about making money." And now, at Sunfire, the company's tall, fair, youthful CEO, Nils Aldag, observes that it is "one of the biggest privileges to be able in your life to work on something which has this level of meaning. That's something which rarely happens to anyone."

Sunfire was founded in 2010, the creation of Aldag, Berninghausen, and a third Dresden-based chemical engineer, Christian von Olshausen. I immediately recognize Berninghausen from an online investor seminar I saw during lockdown: lean, tending to late middle-aged, and wielding an impressive pair of spectacles that snap in half at the bridge, which is magnetic, allowing him to click the two halves together again under his chin, where they hang from a cord until he needs them, a neat maneuver to divert negotiating partners and perhaps grandchildren, and which, to my delight, he executes early in our meeting.

Sunfire quickly acquired another Dresden-based company with electrolysis know-how, a relationship that was consummated when Christian von Olshausen exchanged two cables in

42   a fuel cell—and, presto, the company had its first electrolyzer. "Power-to-liquids," as synthetically produced liquid hydrocarbons are known, are poised to dominate large sectors of the economy. "If you imagine the world in 2050 when hopefully we will be climate neutral," Aldag says from his end of the conference table, "a certain amount of things we're doing today with coal or oil or gas will be done with electricity from solar and wind. Like driving passenger cars, maybe like heating our houses. But in 2050 there will still be a substantial portion of energy which we don't use in the form of electrons because you simply can't electrify those processes. This is the steel industry. This is the shipping industry, the chemical industry. This is the aviation industry."

Sunfire is in this respect like Climeworks: another of those no-more-than-adequately-capitalized startups whose leadership teams confidently predict exponential growth over the coming twenty years. And there is justification for their optimism, namely that they have patented technology that can be significantly scaled and that humanity has no logical choice but to ramp up action against climate change. So it's with a sense of anticipation that I follow Berninghausen outside to inspect the demonstration plant that brings together all the steps necessary to make e-fuel, with the exception of Climeworks's carbon capture technology (the carbon used in the plant is brought in by truck from local plants). Here, using solid oxide electrolyzers, plates of metal and ceramic bonded by glass, water, and carbon dioxide are turned into syngas before being put through the Fischer-Tropsch synthesis. Berninghausen walks me through floor after floor of racks and tubes, pipes, tanks, valves, and ducts wrapped in insulating foil. Only a soft whirr tells me that,

at this moment, small quantities of synthetic fuel are being produced around me.

"The Fischer-Tropsch product comes out liquid and hot," Berninghausen explains as we walk. "It's 240 degrees. It contains the entire portfolio of hydrocarbons, from naphtha, or light gasoline, all the way to waxes, which would normally be in crude oil. Crude contains basically the same elements as our product. But it also contains a lot of dirt. If you look at crude, it's a brown, sludgelike liquid. Ours contains the same valuable products, but without asphalt, without dirt, without any unwanted sulfur and other products that pollute the air. Other than that, it has all the goodies you need for building fuel."

As Berninghausen speaks I experience a thrill of excitement that it is possible in a few hours to do what has taken the earth millions of years. Correction: it is possible to do it cleaner and better. Berninghausen holds up a bottle containing a colorless liquid and unscrews the lid. "This is the future," he says, holding the open bottle under my nose.

"It smells like paint," I say.

"With a hint of fruit," he corrects me with a smile.

A few months after I visited Sunfire, in February 2022, Norsk e-Fuel announced the construction of its first e-fuels plant, in Mosjøen, a town on the Vefsnfiorden in northern Norway. Here, in 2024, technology developed by Climeworks and Sunfire and fueled by thermal power will start producing 12.5 million liters of synthetic fuel a year, rising to 25 million liters in 2026.

According to Waypoint 2050, a report into the future of aviation that was put out by the Air Transport Action Group, whose members include airlines, airports, and plane manufacturers, as many as 7,000 SAF plants may be needed by the middle of this century if the industry is to achieve net-zero.

In theory, since the feedstock for e-fuel is the air, supply is unlimited. In practice, there is no way that all, or even most, of the projected 7,000 plants will produce e-fuels. Growth will be slow for the fairly generic reasons that Gebald outlined for me in Zurich, reasons to do with execution, funding, and the supply chain. It will also be slow because of the bottlenecks that arise when everyone tries to do the same thing at the same time. Even if investment and customers flood in, e-fuel plants require a lot of sustainable electricity and electrolysis capacity—both of which are also in demand from other decarbonizing sectors. And the technology, at least in the early days, will be expensive. E-fuels, whether the technology is developed by Sunfire or the other early players in the industry, which include Siemens and Neste, are proof that the greenest technologies are the most arduous and expensive to achieve. Clearly, if SAF is eventually to lead the way in aviation's efforts to decarbonize, less demanding—less virtuous—varieties need to make the early running.

Dozens of feedstocks can be turned into SAF. They include plant oils, algae, greases, fats, municipal rubbish, sugars, and alcohols. In September 2021, a British Airways Airbus A320neo flew from London to Glasgow powered by 35 percent SAF made from recycled cooking oil blended with 65 percent traditional jet fuel. Dallas Fort Worth Airport has won attention for converting used cooking oil to jet fuel. These are nice stories that make us feel good. But cooking fat won't ever contribute more than symbolically to the decarbonization of aviation. Each month, Dallas

Fort Worth converts just fifteen tons of the stuff, enough to keep an A380 aloft for half an hour. Clearly people aren't eating nearly enough burgers and fries to save the world.

Biofuels are liquid fuels produced from crops, such as biodiesel produced from soya and ethanol made from fermented maize or sugarcane. They've been presented by advocates as a silver bullet for reducing global greenhouse gas emissions, but critics argue that the clearing of native vegetation to make way for biofuel plantations, and the carbon emissions associated with land-use change, can exceed the emissions savings gained by avoiding fossil fuels. Within biofuels, some feedstocks are more promising than others. According to Waypoint, woody biomass, which essentially means thinnings and residue left over from the lumber industry, could provide 57 million tons of SAF globally per year, more than waste oils, municipal solid waste, and industrial off gases put together. At Natchez, Mississippi, in the US's pine-covered southeast, the company Velocys is building a plant called Bayou Fuels that will do just that.

The company's vice president is Jeff McDaniels. "What we're doing at this facility," he tells me when we meet by Zoom, "is in some ways the exact opposite of the way we produce fuels today. Normally we pull hydrocarbons from deep underground. We convert the crude into finished fuels. We combust those fuels and release the CO_2 into the atmosphere. With the Velocys process, trees serve as our CO_2 collection devices. They capture the CO_2 from the atmosphere, they convert it into carbon, and then we move the carbon to our plant. Some of that carbon gets converted into fuel, which is released in kind of a cycle. But an awful lot of the carbon gets converted into CO_2 and is permanently sequestered."

46   Woody biomass is almost equal parts carbon and oxygen. Using solar energy, Velocys's plan is to gasify the biomass, remove CO_2 from the resulting syngas, and build fuel using the Fischer-Tropsch synthesis. The process will run on solar energy and the CO_2 that is removed from the syngas will be permanently sequestered in the saline aquifers of the southeastern United States.

A given technology's carbon intensity is a measure of emissions of CO_2 equivalent, over the lifecycle, compared to the megajoules of energy expended. Negative carbon intensity is a good thing; positive is bad. The fuel that Velocys produces will be carbon negative—the net process has a carbon intensity of minus 144, while fossil-based jet fuel scores positive 90. "One of the most promising things about our technology," McDaniels tells me, "is that we're able to combine the use of sustainable biomass, renewable power, and carbon capture and sequestration to deliver a fuel that offers quite a bang for your buck, so to speak, in terms of carbon savings per gallon. What that means for those who buy our fuel is that for every gallon of fuel they purchase, they'll get four times the carbon savings they would from a conventional SAF that's available on the market today."

The Texas-based low-cost airline Southwest has signed up to buy two-thirds of Bayou Fuel's production. IAG, the parent company of British Airways, will take the rest. Like Norsk e-fuel, Velocys is a technology provider. Its objective at Bayou is to prove the concept and then, as McDaniels puts it, watch bigger companies "come in and use their balance sheets and expertise to make more of these projects happen." But a lot of the drawbacks that are associated with e-fuels also apply to the processes that will be employed at Bayou. Costs will be high, at

least to begin with, and the commercial risks considerable. The
world of renewables is littered with the corpses of plants that
tried and failed to turn cellulosic material into fuel.

That Velocys will be able to make SAF and achieve such
an impressive carbon intensity score is down to a still-rare
combination of conditions. At best, making fuel from woody
biomass—or from municipal rubbish, as Velocys is also planning to do, at a plant it is building in northeast England—will
only account for a small fraction of the SAF that the industry
needs. And the technology will take years to scale, while aviation needs to cut its carbon footprint now. This is the logic that
leads us unerringly to the fastest, cheapest, and easily the most
controversial means of producing SAF. Which is to say, to use
God's good earth to grow it.

There are three reasons why environmentalists hate the idea of
growing crops in order to move humans around. The crops are
often grown in poor countries at the behest of rich ones. The
process is likely to involve cutting down trees. And the primary
function of agricultural land is to produce food.

The recent past shows how badly things can go wrong. In
the first decade of this century, the EU and the US vigorously
promoted the use of biofuels for ground transport. Almost
overnight, a huge market was created for oils derived from soya,
rape, sunflower, and (especially) palm oil. The result was massive deforestation, especially in Indonesia and Malaysia, and
emissions on a vast scale. In the case of US corn ethanol, which
is mixed in huge quantities into gasoline sold at US pumps,
recent research funded by the US Department of Energy concluded that it might be a bigger contributor to global warming

48 than regular gasoline, thanks to emissions resulting from land use changes, along with processing and combustion.

Airlines have come to realize that their reputations suffer if they are associated with such depredations. It's not a good look if you're filling up with fuel from Jatropha grown on former forest land in a country whose inhabitants are hungry. Having said that, if it can be demonstrated that the biofuels are sustainably grown, on land that hasn't been cleared of forest or that is agriculturally unproductive, they're an obvious bridge to net-zero aviation—even if more environmentally supportive alternative uses for the land exist.

In February 2022 I drove through the forests and lakes of eastern Georgia to visit a sustainable fuels producer called LanzaJet near the small town of Soperton. My contact was the company's head of corporate and government affairs, Alex Menotti. For years Menotti lobbied on behalf of Neste, a Finnish oil company that has reinvented itself as a producer of green fuels from waste oils, fats, and greases. Now he does much the same for LanzaJet, trying to convince legislators and regulators to care deeply about biofuels. It can be an uphill task; environmentalists and some regulators—especially the European ones—remain deeply suspicious of biofuels' green credentials. Sitting in the company's office with a group of other new recruits, Menotti replies with a well-worn phrase that at least conveys the sense of going in the right direction: "Let's not allow the perfect to be the enemy of the good."

At the industrial facility that the company shares with its sister company, LanzaTech, which uses a fermentation process to convert waste gases into fuels and chemicals, LanzaJet will process ethanol, the active ingredient in alcohol, into SAF.

Not American corn ethanol, but low-carbon intensity Brazilian sugarcane ethanol. "It's a perennial crop," Menotti assured me, "very efficient, with a very good lifecycle [emissions] score." Gesturing toward several acres of empty scrub behind a link fence, he went on, "If you come back next year, this ground will be occupied by our facility. The ethanol will be brought here by truck after coming ashore at Savannah and go into tank storage and then into the facility to be converted into ethylene. And the ethylene will be converted into SAF."

The process that turns ethanol into fuel is called alcohol-to-jet and it consists of three catalytic reactions. First you remove the water from the ethanol, which produces gaseous ethylene. Then, through oligomerization, you build up the molecules into a longer carbon chain olefin. Then you add hydrogen, which turns the olefins into a mixture of synthetic paraffins. Finally, the product is fractionated to isolate a kerosene that, according to LanzaJet, has superior properties to conventional jet fuel.

LanzaJet has yet to generate any revenue, but it's received plenty of money—from Microsoft's Climate Innovation Fund; from the US Department of Energy—to help it get going. British Airways, Mitsui, All Nippon Airways, and Shell are investors. And while the plant at Soperton will essentially be a large demonstrator, producing a mere 10 million gallons of fuel a year, the company has also announced plans to build a much bigger facility in Illinois, this one capable of producing 120 million gallons annually. The Illinois plant will employ on-site carbon capture and sequestration and power itself through renewable energy, resulting, the company says, in a lifecycle greenhouse gas reduction of more than 70 percent compared to conventional jet fuel.

50 What LanzaJet really has its eye on is the huge quantities of ethanol that are produced every year for US ground transportation, whose market is expected to shrink rapidly with the rapid electrification of cars. "Last year the US made 17 billion gallons of ethanol," Menotti said. "If we were to take that 17 billion gallons of ethanol and put it into jet fuel, we would be able to meet 50 percent of US aviation demand using SAF." Making fuel from US ethanol can be made a lot greener than it is. "We can use green hydrogen, we can use solar electricity, we can use renewable natural gas. The best thing we can do is drive advanced technologies to lower the carbon footprint of the feedstock ethanol."

Menotti's argument that growing corn for fuel can be made much greener is shared by researchers at the Argonne National Laboratory, the Department of Energy's science and engineering research center, which recently found that a combination of sustainable farming practices (such as increasing carbon in the soil and using nitrogen fertilizer more sparingly), using gas generated from animal waste, and sequestering CO_2, could lead to a reduction of emissions to 44.8 grams of CO_2 equivalent per megajoule, which is 153 percent lower than the figure for petroleum jet fuel.

Wandering around the compound with Menotti and with Charlie O'Brien, the plant's bearded, genial, solidly constructed head of safety, it occurred to me that for the first time since I started my research that I was learning about a process that can be scaled quickly. The feedstock is abundant. The technology is well known. If you're a plane manufacturer you only need to ensure that your craft can handle ever-higher proportions of the stuff and policymakers and the market will do the

rest. This is the route that Boeing intends to follow, as evidenced by its announcement in 2021 that by 2030 it will be
manufacturing commercial aircraft capable of using 100 percent biofuel. In sum, SAF is the closest thing to a sure bet in aviation's transition.

The challenge is to slash the process's carbon-intensity.
And this—if it is possible—will have a cascading effect. That
said, there will never be enough SAF to handle even current passenger numbers, let alone the millions more the industry antic-
ipates. And, given that the production of most SAF leads to
some emissions, it's at best an interim arrangement pending the
development of other zero-carbon or true-zero technologies.

As we walked around the site, a large gash through the pine
forests, I asked Charlie O'Brien about potential mishaps. "It
could be a number of things," he replied. "A process pipe that
fails, a gasket leak, things of that nature. But we're going to have
a highly integrated fire protection system. And a gas detection
system that will detect for any gases that are released. Now, if
worse comes to worst, and there is a fire, we'll have fire equipment that stays around the facility."

O'Brien might be held up as the embodiment of aviation's
transition. He spent the best part of twenty years working in
the oil and gas industry in Houston before moving to Georgia
and joining LanzaJet. His family is settling in nicely, he fishes
the lakes for rainbow trout, and he is featured on the front
page of a recent edition of the *TriCounty Connector* (incorpo-
rating the *Montgomery Monitor*, the *Soperton News*, and the
Wheeler County Eagle), above an article announcing the creation of twenty entry-level jobs at LanzaJet. So far things seem

52 to be working out well, but, I asked, to up sticks and leave such an established industry for an entirely new one must make him something of an outlier.

"In Houston," he said with a nod, "they're born into this big petrochemicals industry, they live it, they work it, and then they retire off of it. So they're very hesitant about SAF, about sustainable fuels. But me, I had an open mind, I wanted to try something new. The only hesitation I had was: Is the technology improving, or is it not? I mean, you're leaving a job after eighteen years, you've got a good pension, and you're leaving that for something that's unproven and then it could go belly-up, and you have to return to start over your career. But the technology here has already been proven. And that made me feel a lot better. And the other gentlemen that you saw in the office earlier, too, they all have oil and gas backgrounds. So there's a wave of us that are coming to see that. And we're coming here to LanzaJet."

Hydrogen's Promise

Hydrogen is the simplest and most abundant element in the universe. Burning one kilogram of hydrogen generates enough energy to drive a car 130 kilometers or to heat a home for two days—that's two and a half times more energy than you get from burning a kilogram of natural gas. And all this without releasing CO_2 into the atmosphere. But hydrogen isn't found on its own. It likes to pair up and its lone electron is easily captured by other molecules to make new substances, like fossil fuels, biomass, or water. And it is a law of thermodynamics that extracting hydrogen from these precursors requires more energy than the hydrogen produced will give back. So why bother?

For decades, no one did. It wasn't the energy in hydrogen but its exceptional lightness that made it the fuel of choice for airship manufacturers in the 1920s and 1930s, when millionaires, explorers, and titled folk liked to lounge on art deco sofas as the Atlantic slid beneath them. The *Graf Zeppelin* and her sister ship, the *Hindenburg*, flew well over 1.5 million kilometers without the loss of a single life—until one blustery day

54 in May 1937 when the *Hindenburg* ignited as it tried to land at Naval Air Station Lakehurst, New Jersey. Footage of the majestic cotton-clad projectile dissolving in a plume of flame spread around the globe, and the catastrophe, which killed thirty-six people, brought the age of the commercial airship to a premature close.

Since then hydrogen has been used for processing oil in refineries, for producing methanol for use in plastics, and for making industrial ammonia. Gaseous hydrogen has also been fed into fuel cells, which convert it into electricity that powers buses and trucks. The hydrogen used for these purposes is mostly "gray," meaning it is derived from oil or natural gas, or "black" (from coal), forms that are far from sustainable.

The *Hindenburg* disaster is second only to the sinking of the *Titanic* in the annals of iconic transport mishaps. The substance that also gave its name to the most lethal bomb in existence has suffered from a branding problem. In recent years, however, hydrogen—sustainable hydrogen, that is—has been rehabilitated because it is increasingly seen as crucial to decarbonizing a host of hard-to-abate sectors, from domestic heating to steelmaking and petrochemicals.

Sustainable hydrogen comes color coded. It's green if you separate it from water using an electrolyzer run on renewable energy—as Sunfire's technology would let you do. If produced by pyrolysis, which means driving it out of methane using very high temperatures, again using renewable energy, it's turquoise. Running an electrolyzer on nuclear power produces pink hydrogen. Whenever you add a color to hydrogen, you add a premium. But that premium is set to come down because an avalanche of investment is headed its way.

In 2021 the Hydrogen Council, an industry consortium, estimated that 359 big projects worth a total of $500 billion were underway globally to develop clean-hydrogen production, hydrogen-distribution facilities, and industrial plants that will use hydrogen for processes that currently use fossil fuels. According to Glenn Llewellyn, Airbus's vice president of zero-emission aircraft, and one of hydrogen's most important advocates, the cost of renewable hydrogen will drop by 30 percent over the current decade and will halve by 2050. Economies on this scale have the power to subdue even the laws of thermodynamics.

So why, in the succession of technologies that are going to decarbonize aviation, is this wondrous molecule only a distant second after SAF? Shouldn't it be first? Hydrogen's tragic flaw is that, although it may be light, it is roomy. In fact it takes four to five times the space occupied by conventional fuel, which doesn't leave much room for passengers or cargo.

At a typical atmospheric pressure and temperature, it's a gas. That gas can be compressed. But to get to within a factor of three of kerosene's performance per liter, you need to liquefy it. For that it needs to be chilled to at least −253 Celsius, which requires bulky cooling systems that use energy you'd prefer to spend propelling the aircraft. Another downside of burning hydrogen is that, while it is zero carbon, it does produce nitrogen oxide and contrails, both of which have a warming effect on the climate.

The alternative to burning hydrogen is to feed it into a fuel cell. This is essentially a back-to-front electrolyzer. A catalyst at the negative electrode, or anode, separates hydrogen molecules into protons and electrons, which take different paths

56 to the positive electrode, the cathode. The protons migrate through the electrolyte to the cathode, where they unite with oxygen and the electrons to produce water and heat. The electrons go through an external circuit, creating electricity that can be used to power a motor spinning a propeller or a fan drive.

The only undesirable product of powering a plane with a hydrogen fuel cell is water vapor that produces contrails. And there are ways of minimizing the effects of contrails, which means that fuel cell hydrogen propulsion can be brought close to true zero.

Whichever option you choose—burning hydrogen or putting it through a fuel cell—in terms of the design obligations it imposes on the planemaker, both methods send you back to the drawing board. Hydrogen has "clean-sheet technology" written all over it.

Supposing that the production of sustainable hydrogen soars the way it is predicted to, and there is enough left over from more vital sectors to keep aviation well supplied, there will be other impediments to its widespread adoption. Hydrogen isn't a like-for-like drop-in like SAF. It may be produced a long way from airports and will need to be piped or trucked in. If it's to be combusted, airports will need to be able to liquefy, store, and handle a substance that's very, very cold, and that is highly inflammable—*Hindenburg*-style inflammable. As Mark Workman, a green energy specialist at Imperial College in London, told me, "Hydrogen requires a wholesale change to the way the airport is configured."

Given its considerable potential, it seems perverse that so far Boeing, which came out of the pandemic suffering from production delays and Long Trump (which is to say, a persistent and debilitating aversion to policies aimed at mitigating climate change), has shown no appetite for it. In 2021, Dave Calhoun, the company's chief executive, said, "My path . . . between now and 2050 will not include the introduction of a hydrogen-powered airplane."

Boeing's skepticism can be contrasted with the enthusiasm of Airbus, which in September 2021 declared its ambition to produce a commercially viable hydrogen airliner by 2035. Airbus is currently considering designs for three different zero-emission planes that would be powered by hydrogen burned in a turbine. One of the three is a 120–200 passenger jet with a range of more than 2,000 nautical miles. A second is a propeller plane capable of taking up to 100 passengers half that distance. Both models are not unusual in appearance but have elongated fuselages for holding liquid hydrogen tanks. The most futuristic of the contenders looks like a bat, or possibly a guitar pick, which is to say that its exceptionally wide body blends into the wings (it's in the roomy interstices that the fuel will be stored). In the words of Glenn Llewellyn, this so-called "Blended-Wing Body" model could be the "ultimate high performance hydrogen aircraft" of the future.

Airbus has given itself until the end of the decade to decide which model to follow, but the company is at pains to stress that it cannot make the transition alone. "The whole topic," Llewellyn believes, "requires a lot of collaboration outside of the aviation industry: with the airports . . . with energy companies looking at the infrastructure to get the hydrogen from its production site to the airport, and with the renewable energy sector." Not forgetting, of course, the predictable need for taxpayer money. "The renewable energy and hydrogen economy," he went on to say, "needs to be supported . . . and then with the scale the costs start to come down."

Llewellyn has popped up on YouTube showing off the repurposed A380 that is now the company's "hydrogen propulsion flight laboratory," with two stubs added to the rear of the fuselage, each one containing a hydrogen-powered turbine, and hermetically sealed tanks in the hollowed-out cabin where liquid hydrogen will be stored at −253 Celsius before being gasified prior to combustion. The gas burns at a much higher temperature than conventional jet fuel, so special cooling and coating materials are also having to be developed.

Skeptics of Airbus's program point out that the vast majority of CO_2 emissions come from medium- and long-haul flights, while Airbus's efforts are aimed at the much less damaging short and regional segment of the market. They also find it suspicious timing that the Airbus announcement of a hydrogen plane came close on the heels of a French government announcement of a 15 billion-euro stimulus package for the aviation industry, including 1.5 billion euros to spur research on a "carbon-neutral plane" to be operational by 2035. But if abatement technology is a mountain, and the toughest categories are at the summit, Airbus is at least active in the foothills. And why should France's active promotion of hydrogen-powered flight be considered any less legitimate than the US government's promotion of SAF? Airbus's chief executive, Guillaume Faury, may not have been exaggerating when he said that the shift to hydrogen aircraft and other alternative power sources will mark "the most important transition this industry has ever seen."

For understandable commercial reasons, Airbus has steered clear of the fuel cell route to hydrogen propulsion. It's arduous and expensive. In 2020, management consultant Roland Berger judged that the hydrogen fuel cell was the decarbonization pathway least adapted to existing aviation technologies, whether that be engines, airframes, or electronics. The overall complexity of developing the hydrogen fuel cell was deemed "very high," while hydrogen combustion was merely "high."

Nonetheless, in 2017, when tech entrepreneur Val Miftakhov formed his hydrogen aviation startup, ZeroAvia, it was fuel cell technology he chose to pursue. For all its uncertainty, the hydrogen fuel cell is not without advantages. First, unlike hydrogen combustion, it has the potential to be true zero—a big draw for green investors. Second, in small airplanes, fuel cell systems are more efficient than turbine combustion systems, and if there's one thing we already know about the coming transition to renewable hydrogen, it's that—to begin with, at least—there won't be enough of the stuff. Any energy-consuming business that extracts that maximum from its main feedstock is simply protecting its bottom line.

Born in Russia, Miftakhov grew up among the oil fields of Western Siberia. "My father got his master's in aviation technology, even though his work was building refineries in new oil fields," he told me when we spoke in May 2021. "I remember talking to him about airplanes and stuff like that when I was ten years old."

In 1997, Miftakhov arrived in the US on a student visa and after completing his doctorate in experimental physics at Princeton he started a company, eMotorWerks, which converted

60   gasoline cars to electric power. "We were inspired by what Elon [Musk] was doing with Tesla," Miftakhov recalled, "and we thought that the faster way to get everybody into electric cars would be to move the installed base to electricity. What we didn't take into account," he went on with a smile, "was that everyone in the Western world likes new cars and the value proposition of 'hey, bring us your old car and we'll make it electric and you have to pay us a ton of money and it will have a shorter range than when it ran on gasoline . . .' was pretty weak."

Miftakhov's career to that point had been on the ground; a holiday to Hawaii from his new base of California rekindled his childhood passion for the air. "I took a helicopter tour while I was in Hawaii. They had ex-Vietnam pilots flying the aircraft and they were doing all kinds of high-performance take-offs, so when I got back to California I enrolled in pilot school right away. After that I flew pretty much around the whole of California over three years, in various helicopters, before getting my fixed-wing license. I've been pretty excited about aviation ever since."

Listening to Miftakhov's story, I found myself sympathizing with the young man, who had recently emerged from a Soviet culture that prized longevity and cannibalized prodigiously now colliding with a society that lives for newness and thinks nothing of throwing things away. But Miftakhov was a quick study, and as soon as he moved eMotorWerks into electric vehicle charging, the business looked up. In 2017, it was bought by Italian energy company Enel for a reported $150 million, and Miftakhov set up ZeroAvia.

ZeroAvia's objective is to make a powertrain that can be dropped into existing airframes and run on green hydrogen—

confounding the naysayers at Roland Berger—and if this
sounds suspiciously like eMotorWerks's first, failed mission,
the business case for conversion is stronger for planes than it
is for cars. Airplanes typically only pay back their purchase cost
after a decade or two of service and any measure that prolongs
their useful lives is welcomed by the airlines. While ZeroAvia's
initial plan is to fit engines into short-hop commuter planes
like the Cessna Caravan and the de Havilland Canada Twin
Otter, its longer-term target is the same regional market that
Airbus hopes to conquer using combustion. "We think we can
get to seventy seats in about five years," Miftakhov told me.
"And by 2030 we think we can get to 100-seat." That would be
five years ahead of the deadline Airbus has given itself for deliv-
ering a commercially available hydrogen combustion plane. On
YouTube, Miftakhov promised "4x lower fuel and engine main-
tenance costs, with no harmful emissions," a vow that has gar-
nered the company more than $100 million in funding from
green investment funds, oil companies, and airlines, along with
a £12.3 million grant from the British government.

What will decide the success or failure of Miftakhov's
endeavor isn't the preference of the airframe manufacturers
but that of their customers: airlines and cargo companies. "We
go first to the operators," Miftakhov told me. "They're the ones
making purchasing decisions on what kind of aircraft, what
kind of engine, what kind of fuel. And then, together with us,
the operators persuade the aircraft manufacturers to go with
this type of propulsion." Conveniently enough, the same operators that might go on to become champions of ZeroAvia's
fuel cell technology—the likes of United, British Airways, and
Amazon—are already investors in the company.

A few months after meeting Miftakhov I was given a tour of ZeroAvia's R&D center at Cotswold Airport, former home to the Royal Air Force's aerobatics team, the Red Arrows, in deepest, greenest Gloucestershire, and I began to appreciate something of the complex technical balances ZeroAvia must contend with. "We have to get the maximum amount of power from the smallest real estate for the lowest mass," a technician told me above the hum of a fuel cell stacked in a modified shipping container, "and you need to size your fuel cell for all the different operating scenarios. So the take-off, for example, is very power intensive, but it only lasts for five to seven minutes. And then you're in a cruise requirement, which needs significantly less than that take-off requirement. So do you size your fuel cell for the peak requirement or the cruise requirement?" Then there are the unknowns associated with running a fuel cell in unfamiliar atmospheric conditions. "Everybody's put a stack in a car and run it around for 10,000 hours," the technician went on. "Now the question is, when you take it to 15,000 feet, how does the stack react?"

In addition to the relationship between the power-density of the fuel cells and their weight, there's the basic challenge of storing liquid hydrogen at −253 Celsius, gasifying it, and feeding it into the fuel cell—all in a nineteen-seater some distance up in the air. Sergei Kiselev, Miftakhov's childhood friend back in Russia who went on to become ZeroAvia's vice president, told me, "The fuel cell is pretty efficient. But that means that for every kilowatt of power you produce one kilowatt of heat. And the heat rejection of the exhaust that comes out of the fuel cell is not as efficient as that on a turbine engine . . . what we're dealing with here is a completely different type of thermal management."

Kiselev took me into a hangar and showed me a twin-prop Dornier 228 that was having one of its turbines replaced with a hydrogen electric engine in preparation for a test flight the following day. Until the technology is proven, the thrust for the Dornier and other test models is provided by a combination of a fuel cell and a lithium-ion battery. "We have an incremental approach," Kiselev told me. Step-by-step "we replace the battery with a fuel cell and only then go to two fuel cell electric engines. Next year we will fly completely on hydrogen."

Whether in Britain, the EU, or the United States, today's aviation innovators all agree that the sudden and simultaneous emergence of innovations in need of certification has given rise to a different, less adversarial approach toward new technologies on the part of the regulators. According to Kiselev, ZeroAvia has a good relationship with Britain's regulator, the Civil Aviation Authority (CAA), but the company's cause suffered a setback on April 29, 2021, when a Piper Malibu Mirage being used as a flying testbed for ZeroAvia's hydrogen propulsion system crash-landed and one of its wings broke off. To the company's relief, an investigation by Britain's Air Accidents Investigation Branch found that the crash wasn't caused by the plane's hydrogen technology but an overzealous safety feature that shouldn't be hard to remedy. The team's errors, according to the AAIB, included insufficient ground testing and ineffective emergency procedures—not, however, a failure in the means of propulsion.

If there's one thing, other than a lack of funds, that will stall the development of a promising aviation novelty, it's the perception that it's unsafe—call it the Comet effect. But ZeroAvia's competitors haven't been put off. In the summer of 2022, the

64   Hydrogen Aircraft Powertrain and Storage System, a consortium of seventeen companies that enjoy backing from the Dutch government, announced plans to power single-aisle passenger planes between the Netherlands and the UK using a hydrogen fuel cell. In California several more startups are working to drop hydrogen fuel cell power trains into existing planes. But there's some way to go before the hydrogen fuel cell becomes a scalable technology capable of having an effect on aviation's emissions. And memories of the *Hindenburg* linger.

The *Hindenburg* disaster didn't quite kill the airship. But it reduced a serious mode of transport to little more than a gimmick. Many of us have at one time or another looked up and seen a blimp—an airship without a rigid structure—drifting overhead with the word "Goodyear" emblazoned along its side. In the 1970s, Pink Floyd sent a massive inflatable pig, Algie by name, high above the London sky for some experimental album artwork (no CGI then!), before a gust of wind tore him from his mooring and he disappeared over the horizon.

   Algie had the right idea. His belly was full of helium, which is second only to hydrogen in lightness and has the advantage of being non-flammable. The *Hindenburg* would also have used helium as a means of lift had it not been for the fact that the US, which enjoyed a monopoly over the gas, had banned its export. Helium exists in commercial quantities in only a few natural gas fields, and has a tendency to escape into space as soon as it is released into the atmosphere. It is important for MRI scanners and rocket engines. And, now—despite its relative scarcity—for a new generation of airship.

In the community of Laruscade, in southwestern France, a factory is rising that will produce a 200-meter-long helium airship for Flying Whales. Belying its English name, Flying Whales is a consortium of French-accented companies, including the industrial gas supplier Air Liquide and the enginemaker Safran, that is backed by the governments of France, Monaco, and Quebec. In the summer of 2022, after completing a 122 million euro investor round (the company's third), Flying Whales's founder, Sebastien Bougon, appeared on French television to vaunt the craft's qualities: "It will load and unload in stationary flight, so it has no terrestrial footprint, and since it will be electric-propelled, no environmental footprint either . . . the only other vehicle in the world that can [load and unload while hovering] is a helicopter, whose payload is five tonnes [5.5 tons] or so, or, in the case of Russian and American [heavy-lift] helicopters, around fifteen tonnes [16.5 tons]. We go to sixty tonnes [66 tons]. So it's a game-changer."

Flying Whales's CGI-heavy promotional videos show a vast white vessel, only distantly reminiscent of the sea mammal after which it is named, high in the sky. Cables drop from the base of the craft, and into its belly go containers from cargo ships on the high seas or massive trees that have been felled in forests that are miles from the nearest road. Whether it is taking turbines to wind farms or mobile hospitals to typhoon-hit islands, it's the potential of airships to access the inaccessible that Flying Whales is keen to emphasize.

An airship, even one this big, costs a lot less to make than an A320neo or a Boeing 737. And the helium it contains should last the full lifetime of the ship's casing, or envelope, though

66 some "topping-up" might be necessary. Another advantage is that take-off, the procedure that uses the most fuel in the case of ordinary craft, and produces the most emissions, happens at a significant discount. With airships, 60 percent of your lift comes from buoyant lift, in other words, from Archimedes's principle that the weight of the fluid displaced—in this case, the air—gives you your upthrust. Much of the rest comes from aerodynamic lift (from air flow) and ballonets inside the hull, which are filled with air and expand and contract according to changes in temperature and altitude.

It is this adaptability to the laws of physics that makes airships climate-friendly, particularly when their means of propulsion isn't burning kerosene. Tom Grundy, CEO of Hybrid Air Vehicles, a UK-based airship manufacturer, says that his company's Airlander airship will eventually be propelled by a hydrogen fuel cell after entering commercial service in a hybrid format in 2025. The Airlander, he predicts, will boast a "ninety percent emissions reduction compared to anything else flying . . . we'll be net-zero by the end of the decade."

Unlike Flying Whales's craft, the Airlander can carry passengers—which is why Air Nostrum, a Spanish regional airline, has reserved ten. Your flight from Madrid to Alicante will take you well over three hours, compared to just over an hour by ordinary plane. But when aviation is slow, like much else, it becomes fun again—as you gaze through the glass floor while the world drifts beneath you. But more than leisure and cruise travel (the Airlander has a maximum range of 4,600 miles and can stay aloft for five days), it seems likely that airships' main application will be the kind of heavy lifting that Flying Whales expects to make up its primary market.

Airships are back, and likely to grow, if not dramatically
enough to have much of an effect on aviation's climate impact.
And doubts over their primary fuel don't only stem from
its scarcity. "If the main driver for a future airship industry
is to reduce aviation CO_2 emissions," reasons Julian Hunt,
a researcher at the International Institute for Applied Systems Analysis, in Austria, "a helium-based airship industry
will have to rely on a functioning oil and gas industry. It does
not make sense." For John-Paul Clarke, professor of aerospace
engineering at the University of Texas at Austin, a return to
hydrogen is the obvious solution. "It can be produced greenly
and more and more cheaply with each passing day . . . [and] we
have learned a lot over the years about how to handle [it]."

Fighting against this logic is the image, ingrained in the
collective memory, of the *Hindenburg* in a sheet of flame. The
irony is that the *Hindenburg* disaster wasn't even the deadliest
in airship history. That honor falls to the USS *Akron*, which
was destroyed in a thunderstorm off the coast of New Jersey on
the morning of April 4, 1933, with the loss of seventy-three on
board, and was fueled by helium.

Flying Electric

On December 16, 2021, a group of men dressed in the sober, branded casual wear of the Silicon Valley startup gathered on the tarmac at an airstrip outside Salinas, in California's salad bowl in Monterey County. In front of them stood a black shiny capsule on three spindly legs, the offspring of a suppository and a golf trolley with a V-tail like a hunchback whale. Its single cross-span wing had four banks of three rotor blades—six at the front and six at the back—which made the sound of a loud hair dryer. As the spectators bobbed nervously from foot to foot, the machine rose into the air, tipped a bow, and hovered for ten seconds or so before coming gently to earth. Everyone cheered and clapped and exchanged slightly standoffish hugs that said, "Hey, we're just workmates." Back in the headquarters of Archer Aviation in Palo Alto, watching events on a huge screen, the rest of the company's employees were on their feet, whooping and whistling.

It was the first test flight for Maker, Archer's version of a new kind of aircraft called an Electric Vertical Take-Off and

Landing Vehicle. This masterpiece of nomenclature should on
no account be attempted when drunk; its acronym, eVTOL, is
also hard to get your mouth around; and consensus is lacking
over whether the "e" should in fact be capitalized. The bet that
significant numbers of investors are making is that eVTOLs, if
that is what they continue to be called, will be big. Three months
before the test flight, on September 17, Archer had merged with
a special purpose acquisition company, which has its own lam-
entable acronym, SPAC, but is known more colloquially (and
informatively) as a blank check company.

With the ringing of the New York Stock Exchange bell that
autumn morning, Archer Aviation (ticker symbol: ACHR) was
$857.6 million richer and free of many of the financial con-
straints that inhibit startups in their early, geeks-in-a-garage
years. When I visited Archer in Palo Alto in the spring of 2022,
it was onboarding engineers at a rate of several a week and preparing to move to bigger headquarters. "This is a really cool
period of time for us," Adam Goldstein, the company's then
joint (now sole) CEO, told me in a large, airy meeting room in the
stark, two-story building that served as the company's headquarters (Archer has since moved the fifteen or so miles to San
Jose, which has the advantage of offering the company's growing
workforce more affordable housing). "It's a different mindset
when you're not sitting there stressed about capital all the time.
You can actually go out and execute."

"Execute" for Goldstein means guiding Maker, the company's demonstrator vehicle, through certification with the Federal Aviation Administration (FAA), a process that can take
years and costs hundreds of millions of dollars. It also means
preparing for mass production (to that end Archer has entered

a partnership with Stellantis, one of the world's biggest car-makers), identifying routes and take-off sites in cooperation with municipal authorities and preparing ordinary people for what may be a turning point in their flying lives—the moment when a plane stops trying to be a train, running scheduled services from point to point and packing in large numbers of people, and becomes a taxi on demand.

Hundreds of companies have entered the well-capitalized world of urban air mobility, which will over the next few years be shaken down to a few dozen genuine contenders. Joby, one of Archer's California rivals, and the German company Lilium have also given themselves a head start through mergers with blank check companies. In January 2022, Boeing put $450 million into Wisk Aero, an aviation startup backed by Google's co-founder Larry Page, and they sued Archer for stealing its aircraft design. Then there is Airbus's urban air mobility program, CityAirbus Next Gen, and the Israeli startup Urban Aeronautics, whose small-format CityHawk is propelled by enclosed rotors and, according to a delightful company statement, "has more in common with the birds who nest upon the rooftops of skyscrapers than with nearly every other eVTOL prototype in existence."

From swapping engines on an old Dornier to turning someone's smelly running shoes into fuel, it cannot be argued that sustainable aviation is especially glamorous. eVTOLs are the exception to that rule. All that bespoke composite bodywork and fly-by-wire technology; the way Maker's rotors sit flat, like adorable baby helicopters, for take-off and landing, but tilt for forward flight; the tantalizing promise of full automation; there is something about pure electric that appeals to the antiseptic, unsooty, untactile aesthetic of our age. And if you peer into the workings of a Maker, you'll see no tangle of guts and greasy, carbon-blackened components, but a neatly stowed battery pack and some cables; the cabin gives off the smell of a brutally sanitized rental car.

eVTOLs first started receiving attention in 2016 when Uber Elevate, the platform's urban air mobility division, published a research paper that presented small aircraft running on renewable electricity as part of a broader endeavor to decarbonize society. Uber Elevate also declared that, through automobile-style mass manufacturing, such aircraft would become "an affordable form of daily transportation for the masses, even less expensive than owning a car." Since the report was written, one of its authors, Nikhil Goel, has joined Archer's advisory board, while Uber Elevate has been bought by Joby. The world of eVTOLs is as inbred as the Hapsburgs.

Another boost came in 2019, when Morgan Stanley predicted that by 2040 the market for autonomous urban aircraft could be worth $1.5 trillion, a forecast that made its way into investor pitches and was even cited in Wisk's legal complaint against Archer. Morgan Stanley later revised this figure down to a mere $1 trillion but stood by its assertion that eVTOLs could have as dramatic an effect on transport as cars did in the early twentieth century and commercial airliners after World War II. "Radical changes to transportation modality," the investment house noted, "don't so much 'cannibalize' the current/prevailing form of transport . . . as totally re-invent and re-scale the size of the market itself, frequently by orders of magnitude."

The use cases for which pioneer eVTOL companies are preparing are many and varied. Some haven't been thought of yet.

72   Archer will sell direct to airlines—as it did in 2021 when United placed an order for $1 billion worth of Archer planes, with an option for half as many again. Its second aim is to become a ride-sharing platform; "taking" an Archer will mean using the app to book a seat in a vehicle leaving in twenty minutes from near your home, a commute that—miraculously—won't require you to sit in traffic for hours. Indeed, if Goldstein's happy vision of people hopping from Hollywood to LAX to catch a flight or from Palm Springs to Vegas for an evening on the slots comes to pass, the opening scene of *La La Land* will be relegated to the status of an anachronism.

   Goldstein showed me a 2020 clip of Jay Merkle, the FAA official in charge of certifying eVTOLs, saying, "Probably the biggest question I get . . . is, 'Is this real? Are they really happening?' Yes, this is more than just hype . . . we have at least six aircraft well along in their type certification, which is the first step in introducing the new aircraft into operation." In 2021 Archer received the type of certification Merkle was referring to, the first step toward full commercial certification.

   What half-amuses, half-frustrates Goldstein is that, regardless of the pronouncements of officials like Merkle, the public remains skeptical of eVTOLs and there is a pervasive belief that any number of obstacles—safety; cost; the sheer implausibility of large numbers of people beetling across the urban skies—will stop them from becoming the all-conquering mode of transport their advocates predict. "The FAA are saying that this is happening," Goldstein said, "and we're screaming at the top of our lungs. But in 2024, when it does happen [by which he means, when eVTOLS are taking passengers from Miami International Airport into the city, or from Palo Alto to San Francisco, using existing helipads] I guarantee you everyone's gonna be like, 'this came out of nowhere!'"

There's something endlessly fascinating about America's record of contributing so handsomely to the climate crisis and then regarding it as a business opportunity with a work-around. No other country has produced an entrepreneur who simultaneously wants to save the climate and flee to Mars. And the atmosphere at Archer certainly isn't moody and apocalyptic, as it can be at some of the more introspective European startups I have visited. It's possibly the effect of the California sun, burning ever brighter, ever more assuredly, but the idea that you can create a new category of aviation using components from around the world and that this might actually be good for the climate has been internalized without demur.

The form of cake-and-eatism which says that no problem of human origin is so great that it cannot be overcome, and that this is a splendid thing because it engages every last atom of human ingenuity and profit-lust, has been known to bring out the cynic in me. (Surely better to avoid creating the problem in the first place?) And yet, meeting Geoff Bower, Archer's unfeasibly youthful chief engineer, who comes to work in a Tesla Model Y and wrote his Stanford doctoral thesis on the way albatrosses fly over the ocean without flapping (they borrow energy from the atmosphere), I get the uncomfortable feeling that he may be ahead of me on this and that the problem isn't about abstractions and guilt and other surrender-monkeyish traps (this as Bower goes to work defiling a perfectly innocent whiteboard with terrifying formulae and equations). It's about physics.

74   The problem is taking off and landing. The bigger the area covered by your rotors—the disk area—the less power you need in order to take off and the more you need in order to cruise, because the drag is higher. The balance that Archer tries to strike between disk area and drag involves tilting Maker's front six rotors in cruise and relying on its fixed wings to provide lift in horizontal flight. (And Maker, it should be recalled, is only a demonstrator vehicle. In November 2022, Archer unveiled its production aircraft, Midnight, which seems to differ from Maker principally in that it has a nose like a sperm whale's.)

But even if you're good at disk area and drag, there's no getting away from the fact that, of all the fuel technologies surveyed in this book, electric batteries offer by far the least power for the most weight. A kilogram of gasoline holds 13 kilowatt-hours of energy. A kilogram of lithium-ion battery holds not even 0.3 kilowatt-hours.

That we're even able to contemplate electric-powered aircraft is thanks to Musk, whose Tesla, Inc., has done more than any other entity, public or private, to stimulate improvements in the lithium-ion battery; that puny-sounding 0.3 kilowatt-hours is in fact five times more energy than the old lead-acid battery could muster. And this is without mentioning the other advantages of electric propulsion. You can distribute energy to different parts of the plane without wasting a lot of it as you do with a combustion engine. Distributed propulsion reduces drag, and electric motors are between two and three times more efficient than combustion engines. Oh, and batteries can be recharged.

Although flying an electric plane produces no old-school CO_2, nitrogen oxide, or water vapor, that doesn't mean they're

good for the climate. The lifecycle climate impact of eVTOLs
depends on other things. The first of these is the impact of the
components that go into them. This can be alleviated through
recycling: batteries continue to have terrestrial applications
long after they are past their shelf life in the air, and in the future
eVTOLs may be built of thermoplastics that can be remolded
many times. Depending on the material that is used to make
the cathode, different batteries degrade and need to be thrown
away less frequently than others, which in turn reduces climate
impact at the point of production. But more than anything else,
the sustainability or otherwise of all-electric craft depends on
where the electricity that charges their batteries comes from.

Just how far electric aviation must progress before it
becomes truly climate friendly was illustrated in a 2019 assess-
ment by environmental and aviation scientists in the UK and
the US. Basing their calculations on the 2015 average US grid
CO_2 intensity of 456 grams of CO_2 per kilowatt-hour, they found
that all-electric aircraft would have a lifecycle carbon intensity
20 percent *higher* than that of their modern jet engine counter-
parts, though taking into account non-CO_2 impacts brought
that figure down by around 30 percent. The International Energy
Agency has forecast that the world's electric power demand will
double between 2010 and 2040 (without accounting for aviation or ground transportation electrification), but that emissions from generating electric power will drop by only about
5 percent due to continued use of coal. In India, taking an electric train can be more damaging to the climate than flying. On
the other hand, if you're running a plane using electricity from
the Brazilian grid, which has a high share of renewables, your
lifecycle emissions are likely to be lower than they are in China,

where most electricity comes from coal. Nothing is truly climate friendly unless everything about it is climate friendly. And this, naturally, isn't something that a plane manufacturer can give a guarantee about.

Speaking to Geoff Bower and his colleagues at Archer, learning about the differences between an aircraft performing under optimum conditions (a full payload, the most cost-effective range, and perfect weather) and conditions that are likely to exist in real life, including the obligation to carry extra energy reserves, I began to appreciate how many variables complicate the development of eVTOLs. I also began to understand that although Archer is an aviation company in fact, it isn't so in spirit. While eVTOLs are regulated as other aircraft are, and are bound by the same physical laws, their competitors aren't other planes. They're cars.

"It's really hard to exceed what an electric automobile does [in terms of energy efficiency]," Bower told me, "because we fly faster. It basically comes down to speed. However, due to the efficiency benefits of electric over internal combustion, we can do much better than I[nternal] C[ombustion] powered cars." For a flight in the twenty- to thirty-mile range, "it's not going to be realistic to beat an electric car. However, we absolutely need to, and will, beat IC cars on an energy-use-per-passenger-seat-mile basis by roughly a factor of two."

A few days after leaving San Francisco I wrote an email to Louise Bristow, the company's British marketing supremo, expressing my appreciation at having been able to interview "so many clever people," to which I might have added, "and so diverse," from which you shouldn't infer that they're not white men (they are, or most of them), but rather that their CVs testify to an impressive variety of experience. Dave Dennison, the company's vice president of engineering, did thirteen years at Bell Helicopter, and Eric Wright, head of certification, was at the FAA (this gamekeeper turned poacher claims kinship with the Wright Brothers), while the company's data man, Jon Petersen, was at the Federal Reserve, Uber Pool, and "a European carrier that had a big problem with crew strikes in France." Bristow, my compatriot (the Mini Cooper she drives has Union Jack rear lights), came to Archer from Tesla, while Bower himself is one of a big cohort of Stanford engineering PhDs who went on to make a career in eVTOLs, in Bower's case, via Airbus's (now discontinued) eVTOL program, Vahana.

These accomplished people are now engaged in an interesting scientific and commercial quest at good rates of pay, pleasantly living out what may be the final decades of an inhabitable planet Earth. And all the while, in the time I spent moving from meeting room to meeting room and then onto the deck to do battle with the frankly inedible chicken salad I was offered for my lunch, or out at Salinas peering under the bonnet of a Maker, all seemed confident that, while Archer faced serious and invigorating competition among from fellow startups like Joby, none of the eVTOL divisions of the established players stood a chance.

As Matt Deal, test flight lead, who also worked at Vahana, put it out at Salinas, "I don't think they're going to be as successful as startups, just because startups move faster. And that's probably my biggest takeaway from my time with Airbus. They just have bureaucracy. They have processes that are built up from a long history of safety in aviation, while startups tend to think more along the lines of 'fail fast, learn your lessons

78 early, get to a product sooner.' And we're leading. Because we're starting to work with the regulator early, instead of operating in the normal paradigm, which is you get all your ducks in a row first, and then you go through certification. Since nobody's done it, there's no ducks to get in a row, we don't know what the row looks like. So I think for that reason the startups are leading."

Toward what, exactly? A commercially viable product or a genuine weapon against climate change? That was the question that nagged at me as I made my way north from Salinas, between polytunnels and vines and neat rows of lettuce and broccoli, back to San Francisco and the airport and home. And a few weeks later when I spoke to Richard Aboulafia, the aviation analyst at AeroDynamic Advisory, he posited a discrepancy between mitigating technologies that are scalable (not very good for the environment), and those that are good for the environment (not very scalable). No prizes for guessing which category Aboulafia puts eVTOLs in—this "bizarre circus sideshow . . . in which everybody has decided that the thing that should be most funded is the thing that does the least good." In Aboulafia's view, Archer and the others are "designing air vehicles that have no application in the world of aviation . . . which is fundamentally the opposite of decarbonization. That's re-carbonization."

Decarbonization doesn't mean creating new demand for emissions, no matter how cleverly their impact is reduced. It means replacement—getting rid of what is costly to the climate and putting in its place something less damaging, or, better still, not damaging at all. And Aboulafia is right to question whether this is what eVTOLs achieve. The money and ingenuity that are going into the sector are indeed stupendous, but the only way that all these resources will help the climate in the wider sense—that's to say, lessen the effects of the wrecking ball that is currently smashing its way through our weather—is for electric flight to assume heavier loads over longer distances and enter the grown-up world of aviation proper.

If we accept that SAF is currently the only pathway to zero carbon long-distance flying, and that hydrogen will take care of the regional segment, that still leaves the 45 percent of flights whose range is under 500 miles. Two electric strategies are possible solutions. The one that is closer to hand, the let-not-the-good-be-the-enemy-of-the-perfect option, is hybrid electric. A hybrid engine involves burning fuel of one kind or another (kerosene if you're feeling villainous, SAF if you're saintly), and using that to charge and/or augment your batteries. In June 2022, Rolls-Royce, the world's second-largest aircraft engine manufacturer, unveiled the prototype of a new compact engine, a "turbogenerator," that burns fuel to either turn propellers directly or charge batteries onboard, enabling aircraft to switch between power sources depending on the phase of the flight (and how much power remains in the batteries). The generator can be scaled to deliver power in the range of 500kW–1,200kW, handily capable of propelling single-aisle airliners well over the 150 miles that is the absolute upper limit of the most ambitious eVTOL. To reduce emissions while using such a configuration, the pilot could switch to battery power during especially carbon-intensive phases of flight, whether that be passing through ice-saturated parts of the atmosphere (to reduce contrails) or during take-off and landing. But emissions there will be. They are unavoidable.

80   The holy grail isn't hybrid but full electrification, and this requires radically improved battery performance. Since the turn of the millennium, frenetic R&D has given lithium-ion batteries, on average, about 4 percent more specific energy every year. From the high-voltage lithium-ion batteries that are currently under development at Rolls-Royce, to the lithium air batteries that Japan's National Institute for Materials Science unveiled in 2022 and the "solid-state" battery that is being pursued by QuantumScape, a US startup, the critical element in electric aviation will carry on getting lighter and more powerful. The only question is how fast.

Night and day, rain or shine, in his laboratory at Pittsburgh's Carnegie Mellon University, robots controlled by Venkat Viswanathan, one of the world's leading battery experts, conduct experiments aimed at achieving breakthroughs in performance. In 2021, Viswanathan and his colleague Shashank Sripad wrote that "a battery-pack specific energy of 800 wh/kg could potentially be reached at around mid-century." In other words, in around 2050 lithium-ion batteries will be capable of powering a single-aisle A320 or Boeing 737 the 600 nautical miles that constitute a regional mission. Furthermore, assuming strong continued progress, all-electric "aircraft with . . . a range of 1,200 nautical miles . . . could replace more than 80 percent of all aircraft departures" sometime in the second half of this century.

Capability on the workbench does not necessarily denote capability in the real world, however. As Alan Epstein, a former Vice President of Technology and Environment at Pratt & Whitney, and a professor emeritus at MIT, put it when we spoke in the summer of 2022, "You have to distinguish between range and mission length. For around 1,400 miles—the actual amount may vary considerably depending on the weather and traffic at the time of departure—the airlines routinely carry more than 50 percent more fuel on board than you need just to fly that distance." This is because in case of unforeseen events, "you have to be able to divert to another airport and often . . . the only airport that you know is open is the one you took off from." If your fuel is kerosene, with its high energy density, excess fuel can be carried without undue trouble. Translated into electric flight, however, the principle of excess capacity plays havoc with your finely calibrated trade-off between battery pack and distance. "So a 500-mile range airplane may only be able to fly 200-mile missions, or 250-mile missions," Epstein said.

Everyone likes an optimistic vision and the truth is that no one actually knows how fast battery technology is going to develop. The two days I spent visiting Archer were glass-half-full days. Then, toward the end of my conversation with Geoff Bower, after he had patiently explained his formulae to me, I felt the familiar, unsettling tug of legacy aviation. "This technical prowess sounds very fine," I said, "but the decarbonization of the aviation industry can be completely upended by passenger numbers. Relatively few people in the developing world have been in an airplane, and understandably many would be interested in trying. What then?"

"That's the long-term hard problem for aviation," one of the world's best aerospace engineers ruefully acknowledged. "I don't know the answer."

Change

"Less than 20 percent of the world's population has ever taken a single flight, believe it or not. This year alone, 100 million people in Asia will fly for the first time." That was Dennis Muilenburg, Boeing's then CEO, in the glory days of 2017. And although COVID landed the industry a heavy blow, what persists is the widespread assumption that most people who haven't yet flown aspire to do so and that high-growth economies such as China represent an almost bottomless wellspring of new passengers.

The People's Republic will in large measure decide the future of aviation, not because it makes excellent aircraft that everyone wants to buy (it doesn't, at least not yet), nor because it's a source of ingenious net-zero aviation technologies (ditto), but because a vast and growing number of its citizens can afford plane tickets and airlines have sprung up to meet demand. According to a paper put out in 2020 by researchers from the China Academy of Civil Aviation Science and Technology and

Nankai University, between 1979 and 2016 the number of flights
taking off from Chinese airports increased by an average of
almost 13 percent a year. IATA's prediction had been that China
would overtake the US as the world's biggest aviation market by
seats in 2024; the uneven impact of COVID brought that mile-
stone forward to 2020.

Where flights lead, emissions follow. In 2018, Chinese aviation produced 105 million tons of CO_2, making up 13 percent
of global aviation emissions, second only to the US contribution of 200 million tons and more than triple the 33 million tons
of CO_2 produced by the third biggest emitter, the UK. And it's
instructive to compare China, a spectacularly emerging aviation
market, with a mature market like the UK, since it illuminates
the wrangles that occur when the global aviation industry tries
to spread the burden of mitigation. The moral and technical
problems come when emissions in one market increase just as
those in another stabilize and start to fall: in other words, when
one market holds out the promise of emissions and another has
them under its belt.

Aviation in the UK did its growing between 1990 and
2018 and is ready to slim. If UK aviation follows the "balanced
net-zero pathway" that is envisaged for it under the carbon
budget drawn up by the government's Climate Change Committee, a combination of continuing improvements in aircraft
efficiency, take-up of SAF (which may account for 8 percent
of fuel in 2035, increasing by a little more than 1 percent each
year), and a new generation of aircraft will more than offset any
further increase in passenger numbers, bringing the country's
annual aviation emissions down to around 25 million tons of

84   CO_2 by 2050. These uneradicated emissions, the committee notes, will need to be addressed through "significant amounts of [greenhouse gas] removals."

Now look at China—getting paunchier fast. According to the Chinese research paper I mentioned above, three scenarios are possible for the country's aviation emissions. Under business as usual, emissions will reach 323 million tons by 2035 and 456 million by 2050. If the sector follows a "development scenario," according to which distances traveled increase, fuel consumption stays constant, and there is no biofuel available, emissions will go up to 351 million tons by 2035 and 516 million tons by 2050. Under a third, "low-carbon" scenario, where biofuel use rises until it accounts for a quarter of fuels by 2050, emissions will reach 269 million tons by 2035, and 310 million tons by 2050.

As evidenced by the country's ambitious plans to decarbonize other sectors of the economy, China is far from blind to the effects of climate change. But commercial aviation is a field that, in Chinese eyes, exemplifies the double standards of Western countries when it comes to denying emerging economies the lift-off potential they themselves enjoyed in the nineteenth and twentieth centuries. "Common but differentiated responsibilities" was coined in 1992 as a formula for distinguishing between historic and future emitters. In the context of aviation it manifests itself in the principle that developed economies like the EU, the UK, and the US, having contributed a disproportionate amount of historical emissions, and even now being the homes of the major aerospace industries, must take extra and early action to address climate change caused by aviation.

In 2010, ICAO agreed to a non-binding target of "keeping the global net carbon emissions from international aviation from 2020 at the same level" (later revised upward in recognition of the fact that 2020 was an extraordinary year). A resolution passed by the same body in 2016 enshrined the "aspirational" goal of achieving a global annual average fuel efficiency improvement of 2 percent between 2021 and 2050. The resolution included the caveat that "emissions may increase due to the expected growth in international air traffic until lower emitting technologies and fuels and other mitigating measures are developed," robbing it of much urgency, as did the voluntary, delayed, and flawed offset scheme—CORSIA—that was introduced in the same year. ICAO also avoided imposing specific obligations on individual states. Instead, the "different circumstances, respective capabilities, and contribution of developing and developed states to the concentration of aviation [greenhouse gas] emissions" would determine how each state "may voluntarily contribute to achieving the global aspirational goals."

A collective endeavor without individual responsibilities; a summons without an enforcer; the absence of meaningful action in ICAO's plan of action are the results not only of years of industry obfuscation, but also of geopolitics. Even the stress on continued energy efficiencies isn't as innocuous as it appears to be; it hits developing aviation markets, with their older, less efficient fleets (often bought secondhand from Western carriers), harder than mature ones. In 2019, China (along with Russia) called CORSIA a "reversion to the law of the jungle, which will make it more difficult for developing countries and emerging economies to participate in international aviation competition and bring additional cost to these countries."

86   Not that the People's Republic, once it eventually joins
CORSIA, will necessarily be contributing offsets of a high stan-
dard. In 2016 two experts on Chinese environmental policy,
Barbara Finamore and Alvin Lin, estimated that of 55 million
tons' worth of carbon emission offsets that had so far been
issued in China, only a fraction, such as small hydroelectric and
solar projects, would qualify as "medium quality" under criteria
developed by the Stockholm Environmental Institute, one of
the world's most respected environmental research organiza-
tions, and virtually none as "high quality" ones. The majority of
China's offsets, including wind farms, rural methane use projects, and large hydroelectric and solar projects, would be considered "low quality" because they would have happened anyway
and are double counted, that is, they form part of the emission
reductions required by the UN from all countries party to the
Paris Agreement.

Even the International Maritime Organization, ICAO's
counterpart in shipping, has adopted a long-term climate
goal involving an absolute CO_2 emissions reduction target for
2050 (without offsetting) that would require cutting accumu-
lated emissions between 60 and 70 percent against business as
usual, along with a goal of complete decarbonization by 2099.
Only in October 2022 did ICAO adopt its aspirational net-zero
target, and "aspirational" means what it says—it comes with
no enforcement mechanism or penalty for laggards. All the
while, the "emissions gap," the difference between where greenhouse gas emissions are likely to be and where the science
indicates they should be in order for temperature rises to be
limited to two degrees, continues to grow. Call it a gulf between
reality and aspiration, call it evidence for the misalignment of

the nation-state and the climate crisis. Or simply call it the
failure—again—of aviation to get its house in order.

In the course of researching this book, I sent out many emails
introducing myself and my project to the world's big aviation
companies. Many of these emails went unanswered. To put it
another way, if *Flying Green* contains no interview with Airbus
or Boeing, Rolls-Royce or United, it's not for lack of trying. And
this, in turn, says something about the industry's defensive,
prickly state of mind. In the golden age—before COVID, before
anyone cared about climate change—journalists only needed
to say, "Tell me the secret of your success," and aviation's big
players would show them the graphs, seat them in First with
a glass of Dom Perignon, and await the inevitably fawning
write-up. That's because flying back then was an unques-
tioned good, an activity it would be perverse to oppose—
Mark Twain's hymn to travel, "fatal to prejudice, bigotry, and
narrow-mindedness," made flesh.

Getting close to the titans is harder nowadays because they
feel bruised and unloved. When activists declare that future
generations will regard those responsible for today's climate
crisis with the same repugnance that used to be reserved for the
perpetrators of historic abominations like the Holocaust, Big
Aviation shifts uneasily in its seat. Then it does what Big Oil,
Big Tobacco, and Big Pharma have taught it to do: it throws dust
in our eyes.

In the summer of 2022, the British carrier Virgin Atlantic
ran an ad featuring exclusively women, LGBTQ+, non-white,
and disabled people boarding and enjoying a blissful flight.
Never mind that staff shortages, strikes, and a heat wave of

88   Saharan intensity had turned the simple act of exiting the country through a British airport into an unpleasant ordeal—the ad's message was that to fly Virgin was not only to travel in comfort but also to be on the right side of history.

That the diverse human beings depicted in the ad were, by flying across the Atlantic, making a disproportionate contribution to the premature death and displacement of poor people and the annihilation of species, to the end of livelihoods and the failure of harvests, and much of this in the global south, of course went unmentioned. But when I saw the ad—as the Heathrow tarmac melted, as the phrase "unseasonable heat" turned into a meaningless anachronism—I became aware that to admire an airline for being inclusive was beside the point, like praising Mussolini for running the trains on time or Nero for his commitment as a musician.

You can tell a lot about an industry by its hiding place. "Could you take the train instead?" asked that now-notorious KLM ad, and the answer is, "Of course we could, but you—your lobbyists, the governments you're in bed with, the growth model you swear by—are stacking up the arguments for us to take the plane." In the spring of 2022, the Dutch advertising watchdog ruled that KLM's claim that its passengers can fly emission-free was "misleading" because the company's offsets would not result in absolute carbon neutrality. KLM also became the object of the first legal case against an airline alleging greenwashing. "KLM has . . . stuck by the false message that it is on the path to more sustainable flying," argued Hiske Arts, a campaigner at the environmental group Fossielvrij NL, one of the plaintiffs. "There is no way it can do this while planning continuous air traffic growth that will fuel the breakdown of our climate." Absent an instant and simultaneous maturation of all the technologies that I have outlined in this book, Arts's words are a simple statement of fact.

According to a recent report by Influence Map, a think tank that gauges the extent to which companies align their actions to Paris-compliant benchmarks, Europe's major airlines and industry groups employ "a two-point strategy . . . to avoid regulation directly addressing their climate emissions"—a strategy, in essence, of rhetorical enthusiasm for climate targets while opposing any practical measure to achieve them. "At a European level," runs the report, "the aviation sector has communicated high-level support for net-zero EU aviation emissions by 2050 while opposing specific national and EU-level climate regulations to help deliver that target." On the global stage, the report goes on, "Industry has lobbied for the CORSIA offsetting scheme to take precedent over policies addressing absolute aviation emissions reductions. At the same time, using the context of the COVID-19 pandemic, industry lobbyists have successfully pushed for the scheme to be watered down further."

Aviation can't even claim that what it's doing is no worse than other economic actors. "While many industrial sectors are in the process of transformation in response to the EU's strengthened climate agenda," the report goes on, "many airlines have initiated extensive, climate-focused PR campaigns to deflect growing concern from governments and the public over the sector's climate footprint."

The industry and its friends like to keep the mood music around emissions melodious and blame-free. The first argument they make is that aviation is responsible for a mere 2 percent of global CO_2 emissions, a claim that camouflages the sector's

90   substantial non-CO_2 emissions and its deliberate planning for dramatic emissions growth in the coming decades. Their second argument is that aviation is being held back from a more rapid adoption of SAF by inadequate government support—a bit rich from an industry that has received some $30 billion in government bailouts since 2019 and is barely taxed. In general the big aviation players are happiest tinkering with flight paths or making more efficient jet engines, refusing to accept that what's needed if the emissions gap is to be made up isn't a gentle fall in emissions growth but an absolute drop in emissions, and that this can only be achieved, pending the emergence of the new technologies, by a reduction in flights and a fall in passenger numbers.

No one put a gun to my head and told me that in order to write this book I must fly more miles in the course of a few weeks than I had over the previous several years. No one is obliged to take their family on a beach holiday or play the world's major golf courses before they die. These are choices people make. And they would be made with greater circumspection if the ticket price reflected the toll exacted on the environment. (Pending that day, people should offset their own flights according to the most exacting criteria—I offset mine through the German non-profit Atmosfair.)

Greening aviation isn't a mystery. From the introduction of VAT on tickets and a fuel tax on kerosene, to tough SAF blending mandates and an end to frequent flyer programs for all but genuinely net-zero flights, a judicious combination of market and regulatory measures would both dampen growth and raise funds to finance green technologies that would in turn enable the industry to bring down emissions and once again expand—only this time on a sustainable basis. And there are no doubt people working away in the big aviation firms who would genuinely like to see this happen, people who are animated by a sense of technology's potential to reinvent the sector as the force for good it used to be, and might yet be again. Perhaps Airbus's own Glenn Llewellyn is such a person, or Grazia Vittadini, Rolls-Royce's chief technology and strategy officer—but even if so it would be unwise to assume that they constitute a majority in an industry that has spent the past twenty-five years, ever since climate change entered the world's consciousness, fighting one of the most effective anti-environmental rearguard actions in corporate history.

It's not technology that's holding up the decarbonization of aviation. It's money. Be sure in the years to come to look beyond the industry's advertised commitment to net-zero by 2050, and rather at the actions it front-ends to achieve that goal and its willingness to submit to the same fiscal and regulatory rules that have proven necessary for the decarbonization of other sectors. Aviation needs to lose its status as an entitled exception. Pound for pound, flight for flight, it's a massive and growing problem and it needs to pay its way.

Captivating

In Switzerland, Climeworks's head of climate policy, Christoph Beuttler, had painted a seductive picture of e-fuels emitting around 85 percent fewer greenhouse gases than fossil fuel kerosene, and of that already quite encouraging figure being tipped into positive territory through direct air capture and storage (DAC+S).

DAC+S is the name for what you're doing when you suck carbon dioxide out of the air and then, rather than recycle it to make soft drinks or sustainable aviation fuels, take it out of the atmosphere forever. From the Deccan in India to parts of Africa, Japan, and the North Atlantic ridge, DAC+S is possible wherever there is young volcanic rock. If you squirt CO_2 dissolved in water into the basalt in Iceland, say, you release metals such as calcium, iron, and magnesium that mineralize your CO_2 and turn it into rock, which fills the pores of the subsurface. This takes a couple of years or so. It's another example of how humans can accelerate a process that nature needs millions of years to achieve, and it's the science behind the most durable and limitless method we have to fix carbon. "For every cubic meter of porous basalt you can store fifty kilos of CO_2," Kari Helgason, head of research and innovation at Carbfix, an Icelandic company whose mission in life is pretty well expressed by its name, told me. "In Iceland alone we can store all the emissions from burning fossil fuels for hundreds and hundreds of years."

Environmentalists abhor carbon capture and utilization (CCU), by which oil companies inject high-pressure CO_2 into wells to bring crude to the surface, as a particularly pernicious form of greenwashing. Partly because it uses similar technology to CCU, partly because of fears that high-emitting sectors might see it as a license to carry on pumping out carbon, DAC+S has for most of its short history also been viewed with suspicion. If anthropogenic emissions grow as rapidly as projected, they will significantly overshoot the IPCC's upper limit if mean global temperatures aren't to rise by more than 1.5 degrees by 2050, and something will have to be done to claw them back. The deployment of DAC+S is "unavoidable if net-zero CO_2 or GHG emissions are to be achieved," is how the IPCC acknowledged this reality in 2022. "Estimated storage time scales vary from decades to centuries for methods that store carbon in vegetation and through soil carbon management, to 10,000 years or more for methods that store carbon in geological formations."

The promise of carbon sequestration in perpetuity is what has persuaded the likes of Microsoft, the Economist Group, and the rock group Coldplay to buy Climeworks's Iceland offsets, and it was in Iceland—as the guest of Carbfix, Climeworks's local carbon storage partner—that I concluded my research for *Flying Green*. While CO_2 capture technology is part of the most optimistic scenario for the greening of aviation, under which

94 the captured CO_2 ends up in e-fuels, it's also part of the most pessimistic scenario, the one that says that net-zero is unattainable by novel aviation technologies alone and that hundreds of gigatons of CO_2 must be withdrawn from circulation. IATA incorporated both extremes when plotting aviation's path to net-zero by 2050; according to the agency's modeling, 65 percent of projected carbon savings will need to come from SAF and 19 percent from carbon capture and offsets.

That Iceland is a suitable setting for carbon burial isn't simply down to its abundance of basalt. The island is also a treasury of renewable energy, a volcanic hot spot at the meeting point of the Eurasian and North American tectonic plates whose stupendous reserves of heat drive the country's electricity turbines and warm its houses.

As I drove northward from Reykjavik one March morning in 2022, steam and volcanic gases rose from fumaroles that had punched smoking fissures through the thick snow on either side of the road. My destination was the Hellisheidi geothermal power plant, which harnesses the thermal energy of a nearby volcano and turns it into electricity and domestic heating. The area around Hellisheidi has attracted green businesses that buy water, heat, and CO_2 from the plant, from vertical farms to producers of algae for use as fish-feed. In Hellisheidi's gift shop I found a group of North American eco-tourists comparing silica mud masks and algae supplements and slaking their thirst by drinking mineral water that had been carbonated using the plant's CO_2.

Carbfix isolates excess CO_2 that is emitted by the power plant and injects it deep underground. Beginning in 2026 the company will receive liquefied CO_2 that has been captured from European factory flues by ship at a terminal on Iceland's southern coast

and sequester it in the basalt. The EU is giving more than \$100 95
million to develop this initiative, which will reach full maturity
in 2031, after which some 3 million tons of CO_2 imported from
European industries will be turned to stone each year.

From a technical and financial point of view it's relatively straightforward to isolate CO_2 from the flue gases of a
power plant; CO_2 constitutes around half of these gases. Cap-
turing it from the air around us, on the other hand, needs more
energy and is less rewarding because CO_2 occurs in such small
quantities—roughly 400 parts per million. When I asked Kari
Helgason to explain the complexities, he asked me to "imagine
that your child has thrown all the LEGO pieces on the floor and
you're supposed to filter out only the pink ones, and there are
millions of pieces of LEGO flying through your room and you're
picking out a few pink ones. Well, it's the same thing with direct
air capture. Once it's in the atmosphere, it's hard to get it back."

Picking out the pink is expensive. DAC+S is perhaps the cost-
liest way of clearing up a mess it would have been far cheaper
to avoid making in the first place. In 2020 the US payments
company Stripe agreed to pay Climeworks \$775 a ton to extract
322 tons of carbon dioxide from the air. On the other hand, if
you were to capture and permanently store all the CO_2 that is
produced by, say, a cement factory in Belgium, you'd be at best
net-zero, whereas what's needed now, in our world of drastic
emissions overshoot, is net negative. And DAC+S isn't entirely
without advantages. It's surer than planting trees. It doesn't
need to be attached to anything or to be built in any particular
place, but can be put anywhere in the world.

The challenges of DAC+S, as with every other climate
mitigation strategy, concern cost and scale. With sufficient

"deployment and innovation," the IEA said in its 2022 report, costs could fall to under $100 a ton by 2030, though the agency stresses that for this to happen "increased public and private support" will be needed. In 2022, Climeworks announced that its most recent funding round had raised $650 million, bringing the total invested in the company to $800 million. Also in 2022 a phalanx of digital and financial companies set up Frontier, a climate fund that plans to purchase $925 million worth of permanent carbon removal from companies that are developing the technology over the next nine years. If investments in the sector carry on growing at this rate, costs will indeed come down and the exponential growth that Gebald spoke of might just be possible.

Hellisheidi is home to Orca, an arrangement of container-sized units, each one containing twelve DAC machines configured without walls in the interests of efficiency. Orca is a more advanced version of the tumble-driers I saw on the roof of the municipal waste plant in Switzerland. The same acid-alkaline reaction isolates CO_2 in a filter from which it is periodically detached by blasts of waste heat and piped into a processing hall, where it's cleaned and compressed. So far, so familiar—only what happens next at Orca is that the CO_2 is sent to a well-head where it is reliquefied and powered deep into the earth.

The landscape around Hellisheidi is dotted with such well-heads. Each one is housed in a many-faceted silver-colored hut that might have been lifted from the props department of *Thunderball*. Escorting me into one of these huts, my guide, Ingvi Gunnarsson, an indefatigably good-natured scientist from Iceland's wild north who is a member of Carbfix's governing board, tapped on a metal pipe and said, "This is the injection valve. The water it sends down is slightly acidic. The bedrock provides the necessary elements to mineralize the gases: calcium, iron, magnesium." And he laughed delightedly as if to say, "You see the works of nature?"

When I visited it, Orca was the biggest carbon capture facility in the world, which isn't saying much. Mammoth, the larger facility that Climeworks started building a few months later, will be able to suck 36,000 tons of CO_2 every year from the air. I asked Gunnarsson to put that figure into context. After a moment's thought he said, "If a car uses ten liters of gasoline per kilometer, and you drive a car 20,000 kilometers in the course of a year, that makes 2,000 liters. That's two tons of gasoline. And that's times three, because it's CO_2. So for a car it's six tons. Divide 36,000 by six and you get 6,000."

Mammoth by name, mouse by nature; the new plant will mitigate the emissions of a mere 6,000 cars. "It's not much," Gunnarsson admitted, "but you have to start somewhere."

"You know the story behind the Blue Lagoon?" Gunnarsson asked. "It's wastewater from the next-door power plant. The plant was built when the environmental protocols weren't as strict as they are now. They spilled the water outside and released it into the lava next to the power plant. To begin with, the water seeped through the lava field and disappeared. But the water is rich in silica and the seepages clogged up because of the silica scaling. So a lagoon started to grow, with this beautiful blue color from the silica, and people started coming to bathe and those with psoriasis and other skin conditions got much better. So an industrial spill ended up a tourist attraction. And the revenues from the Blue Lagoon—from bathing in the spent fluid—are much higher than they are from the power plant."

It's not much but you have to start somewhere. It could be a summary of aviation's attempts to decarbonize, I thought while negotiating my way between bodies in the Blue Lagoon. And while I had learned from Climeworks that parts of the aviation sector—courier companies, airlines, and travel companies—had inquired about buying carbon credits, as yet there were no announcements. Financial services, yes, the media—both digital and legacy—yes, but not yet the economic sector that most urgently needs to up its game. Where is aviation in the story of DAC+S? Still transfixed by passenger numbers? Still unconvinced by the urgency of climate change? Still evading responsibility for its emissions?

In *Flying Green* I've tried to introduce the various pathways to green aviation. All must come to pass if the goal is to be achieved. This process would go a lot better if we all lived under a global system of carbon taxes, but at present only about 20 percent of global emissions are covered by a carbon price, with the result that the global average price is pitifully small, just $6 per ton. The companies that make up international aviation have come some way toward developing the technologies that are needed to make significant dents in emissions. Now they must be forced to accelerate those technologies even if it means a rise in prices, even if it means fewer people up in the sky. *Love it or hate it,* I thought to myself while drifting in the milky waters and wondering pleasantly how you spell "psoriasis," what aviation needs is a session with the thumbscrews.

ACKNOWLEDGMENTS   99

My thanks go to the following for their expertise and encour-
agement: Werner Antweiler, Andrew Newell, David Wolf, Stefan
Gossling, Mirabelle Muuls, Chris Morris, Mark Workman, Mark
Ellingham, Paul Lashmar, Stuart McLaren, Ajay Gambhir, Rupert
de Borchgrave, Tim Johnson, Cait Hewitt, Andreas Pelotti, and
Barnaby Rogerson. Thanks also to the splendid team at Columbia
Global Reports: Camille McDuffie, Nicholas Lemann, Allison
Finkel, and my peerlessly patient editor, Jimmy So.

There's not a lot of gripping reading out there about the various paths to green aviation, but Jeff Holden and Nikhil Goel's "Fast-Forwarding to a Future of On-Demand Urban Air Transportation," *Uber Elevate*, October 27, 2016, at least has the merit of being a historically significant primary source; it was the first document to lay out the potential of eVTOLs.

The history of aviation is, of course, well served; David McCullough's *The Wright Brothers* and Scott Berg's *Lindbergh* are useful biographies of two of the founding fathers, while Lowell Thomas's *European Skyways*, Hermann Geiger's *Alpine Skies*, and a memorable turbulence-and-vomit scene in Sinclair Lewis's *Dodsworth* give a good idea of the thrill, ambition, and physical discomfort that were essential to the sector in its youth.

For anyone who wonders how hydrogen can help green aviation, Marco Alvera's *The Hydrogen Revolution* is a good way in, while *The Global Airline Industry*, edited by Peter Belobaba, Amedeo Odoni, and Cynthia Barnhart, is packed with market- and operations-related minutiae regarding the industry at the peak of its trajectory.

Alastair Gordon's *Naked Airport* and Christopher Schaberg's *The End of Airports* are enjoyable romps through the culture of being on either side of airborne, while the US House of Representatives' Committee on Transportation and Infrastructure's *Final Committee Report: The Design, Development and Certification of the Boeing 737 MAX* lays out in painful detail the unhealthily close relationship that exists between the industry and its regulator.

No list of aviation books is complete without mentioning that epic but defunct genre, the airport novel, of which Arthur Hailey and Jackie Collins were the pinnacle, or the nadir, depending on your perspective, and are now breaking down in a landfill somewhere near you.

NOTES

INTRODUCTION

11 **"We shall not need to fuss":**
David McCullough, *The Wright Brothers* (Simon & Schuster, 2016), p. 120.

11 **"the valleys of clouds":**
Hermann Geiger, *Der Gletscherflieger* (Universitas-Verlag, 1961), p. 19.

13 **Subsidies doubled for airmail:**
Carl Solberg, *Conquest of the Skies: A History of Commercial Aviation in America* (Little Brown, 1979), p. 64.

13 **"Helpless as dice in a box":**
Sinclair Lewis, *Dodsworth*, p. 259.

14 **The Lindbergh effect:** Scott Berg, *Lindbergh* (Berkley, 1999), p. 171.

14 **US airlines were flying more passengers:** Solberg, p. 126.

14 **United, TWA, and American would buy their products:** Solberg, p. 146.

18 **the cost of flying dropped dramatically:** Derek Thompson, "How Airline Prices Fell 50 Percent in 30 Years (And Why Nobody Noticed)," *The Atlantic*, February 28, 2013.

18 **cost you around $300:**
Jack Houston, "Why Air Travel Is So Cheap," *Business Insider*, November 8, 2019, https://www.businessinsider.com/why-air-travel-is-so-cheap-2019-11?r=US&IR=T.

18 **airline industry was in the red every single year:** Derek Thompson, "How Airline Prices Fell 50 Percent in 30 Years (And Why Nobody Noticed)."

19 **growth of air traffic has averaged 4.4 percent a year:**
"The World of Air Transport in 2018," International Civil Aviation Organization, n.d., https://www.icao.int/annual-report-2018/Pages/the-world-of-air-transport-in-2018.aspx.

19 **a tiny number of rich people who fly very often:** UK household data show that between 2006 and 2012, 3.4 percent of respondents accounted for 30 percent of all flights, with five or more international flights a year. A Swedish airport survey concluded that the 3.7 percent of the most frequent fliers accounted for 28.3 percent of all flights taken.

20 **just 2 percent of the world's population:** Stefan Gossling and Andreas Humpe, "The Global Scale, Distribution and Growth of Aviation: Implications for Climate Change," *Global Environmental Change*, Vol. 65, November 2020, 102194.

20 **"large vessels will go out of fashion":** Christopher Schaberg, *The End of Airports* (Bloomsbury Academic, 2015).

21 **parents terrified of upsetting their young:** Two episodes give an idea of the clammy relationship

102   between the plane-making duopoly and the authorities that are supposed to stand impartially above them. In 2020, the US Congress's Committee on Transportation and Infrastructure found that Boeing had used its influence over the Federal Aviation Authority to expedite certification of the 737 MAX, leading to two catastrophic crashes with huge cost of life. The committee found "multiple . . . examples where FAA management overruled a determination of the FAA's own technical experts at the behest of Boeing." Notwithstanding this and other damning findings, the US Justice Department fined the company a nugatory $244 million for its misdeeds, failed to initiate criminal proceedings, and continued to facilitate its access to billions of dollars in private sector credit. In another incident, this one involving Airbus, emails procured by the European NGO Transport and Environment showed that in 2017 the European Commission invited Airbus to help set the emissions rules by which its aircraft would have to abide, and that "red lines" laid out by the company influenced these rules, https://www.transportenvironment.org/discover/emails-show-airbus-writes-aircraft-co2-rules-commission-france-germany-and-spain-complicit/.

21   **"Continued political support and economic investment will be needed":** "Aviation Benefit Report," The Industry High Level Group, 2019.

22   **11 percent of the world's population who fly in a single year:** Gossling et al (2020).

22   **1 percent who account for more than half the total emissions:** In 2018, CNN produced a perky story about the phenomenon of the "mileage run," a flight or series of flights undertaken for no reason other than to earn miles and bump up the customer's frequent flier status to get perks like free upgrades and priority boarding. "Paul Bevan," the network reported, "flew from New York to Detroit to Tokyo to Singapore and back the same way without ever leaving an airport. The whole thing took sixty-three hours. Bevan says he 'showered, ate a fabulous Indian meal at the food court at 3:00 a.m., mooched around a bit, and flew back at 6:00 for 21,000 MQM's [Medallion Qualifying Miles] and ensured making Diamond [United's top tier, achievable only by spending $15,000 on plane tickets] again this year.'" Climate change wasn't mentioned in CNN's report, which is a bit like reviewing a restaurant in Berlin in 1945 without mentioning World War II.

22   **30 to 35 kilometers per liter of fuel per passenger:** Jo Hermans, "The Challenge of Energy-Efficient Transportation," *MRS Energy and*

Sustainability: A Review Journal, 2017, p. 4.

23 **In first class, it's off the scale:** A Singapore Airlines A380 cabin can carry 471 passengers, with 12 first-class suites requiring about the same space as 60 business class seats. Together, the 72 passengers flying in the top two classes use the same space as 399 passengers in economy. Premium classes use an average of 5.5 times more energy than economy passengers.

23 **You and I call it kerosene:** In a classical refinery, jet fuel (or kerosene) is the middle distillate making up to 10 percent of the crude oil fraction while the majority is gasoline and diesel. Jet fuel is preferable to gasoline for planes as it is less volatile and denser; compared to diesel, jet fuel is lighter and less prone to wax at low temperatures (Frontiers in Energy Research).

24 **3.5 percent of the warming impact:** David Simon Lee, et al., "The Contribution of Global Aviation to Anthropogenic Climate Forcing for 2000 to 2018," *Atmospheric Environment* (Oxford, England : 1994), Vol. 244 (2021), 117834.

24 **have fun at the other end:** Before the pandemic, business travel accounted for 12 percent of airline tickets but 75 percent of revenues on certain flights. In the four years she was US Secretary of State, Hillary

Clinton reportedly flew 1,539,712 kilometers on official business. That's the equivalent of thirty-eight circumnavigations of the earth, much of it accomplished on her personal Boeing 757, which gives Clinton one of the more impressive carbon footprints of all time.

24 **100 times more CO2-equivalent per hour:** Milan Klower, "Quantifying Aviation's Contribution to Global Warming," *Environmental Research Letters*, Vol. 16, No. 10, November 4, 2021, 104027.

24 **"quickest and cheapest way to warm the planet":** Stefan Gossling, an aviation expert who is a professor at the Linnaeus University School of Business and Economics and at Lund University's Department of Service Management, and Andreas Humpe, a tourism professor at Munich University of Applied Sciences, estimate that 1 percent of the world population is responsible for 50 percent of emissions from all air travel (Gossling et al., 2020).

25 **aviation is even further behind:** "Clean Skies for Tomorrow Sustainable Aviation Fuels as a Pathway to Net-Zero Aviation," Insight Report, World Economic Forum, November 2020.

25 **half a percent of the savings:** The study found that eating a plant-based diet saves about 0.8 tons of CO_2 equivalent per year. Each return

104   transatlantic flight avoided saves
1.6 tons. Living car-free for a year
saves 2.4 tons. But governments'
advice on climate change tends to
concentrate on persuading people
to change their lifestyles in ways
that have little impact. Hence
the dominant rhetorical position
enjoyed by recycling. The truth is
that, for an American family, having
one fewer child would deliver the
same level of emissions reduction
as 684 teenagers recycling
religiously for the rest of their lives.

29   **domestic air travel fell by 12.3
percent:** Gossling et al., 2020.

30   **1.50 euro carbon ticket tax:**
In 2022 environmental groups sued
KLM for misleading the public with
its declared "commitment to taking
a leading role in creating a more
sustainable future for aviation,"
arguing that the airline's plans to
fly more and more passengers were
incompatible with the urgent action
necessary to secure a livable future.

30   **the frequent flier, is the
worst polluter of all:** The industry
comes together reliably to quash
any attempt to penalize frequent
fliers under "polluter pays." In
2019, Airlines UK, whose members
include British Airways, Ryanair,
and Virgin Atlantic, reacted to a
proposed frequent flyer levy by
warning that it would "price
people out of air travel" and
"damage the UK's reputation
internationally."

31   **at most 21.6 percent of their
emissions:** As an incentive to
reduce emissions, the offsets
currently being purchased by the
airlines are ineffective because they
are so cheap. The price that airlines
like EasyJet pay in the "voluntary
market" is $6 per ton of CO_2, just a
fraction of the carbon price in the
EU's emissions-trading scheme,
which represents a negligible
additional expense, https://
medium.com/ouishare
-connecting-the-collaborative
-economy/does-easyjet-really
-offset-all-of-its-carbon
-emissions-e71d7371d274.

CHAPTER ONE

34   **with respect to contrails:**
Figuring out the climate impact
of jet fuel burned in the upper
atmosphere is complicated by
the fact that CO_2 can survive
for hundreds of years in the
atmosphere while contrail cirrus
often disappears after just a few
minutes. On the other hand, if
conditions are right, contrails from
a single plane can last for many
hours and expand horizontally to
cover hundreds of square miles.

37   **SAF is a "drop-in":** The
blending rate is currently limited
to 50 percent for safety reasons. As
Alex Menotti of the biofuels startup
LanzaJet explained, "Essentially
planes get addicted to the sulfur in
the fuel. That creates a seal swell in
the fuel system. There's a concern

that if you went to 100 percent SAF [thereby eliminating the sulfur] you might have some brittleness and some leaking because sulfur is a lubricant. So there's more work to be done to test every piece of equipment. But over time I'm confident that we will get to 100 percent SAF."

39 Fischer-Tropsch synthesis: Named after the chemists Franz Fischer and Hans Tropsch, the Fischer-Tropsch process is a series of chemical reactions that uses metal catalysts and high pressures to convert a mixture of carbon monoxide and hydrogen into liquid hydrocarbons. In World War II the German war machine used it to turn coal into fuel. Nowadays the process is being developed as a way of producing carbon-neutral liquid fuels from atmospheric CO_2 and hydrogen.

39–41 "of course we all want to have breakfast": The way that Berninghausen explains the phenomenon of carbon neutrality—or, to put it another way, the circular carbon economy— raises interesting questions about responsibility and benefit. "It's a circular process," he said. "We catch the carbon dioxide and we release it, we catch it, we release it. But the underlying truth is that we leave the oil in the ground and we don't bring out additional carbon, which is now stored underground. And wherever we get the carbon from,

it's not our responsibility that this carbon is there. And I don't know who put it into the air. And also Climeworks doesn't know. We just take it from the air and we release it into the air."

43 According to Waypoint 2050: According to the Waypoint report, the new facilities will typically be built close to feedstock sources and require between $1 and $1.5 trillion in investment. Per year this represents approximately 6 percent of the world's historical annual oil and gas capital expenditure. "The level of investment," the report continues, "will be increasingly achievable as additional specialized producers enter the market, fossil infrastructure is increasingly available for retrofitting, and investors look to decarbonize portfolios."

48 grown on former forest land in a country whose inhabitants are hungry: In the words of a recent study by Princeton University, "Although the plant growth that eventually becomes biofuels absorbs carbon from the air, it takes land to grow plants. The cost of devoting land to bioenergy is therefore the lost use of that land for other purposes. They include directly storing carbon in existing or new forests or producing food, increasing the capacity to preserve or restore forests and other habitats elsewhere while meeting rising food demands. The carbon-neutral

106   assumption in effect treats land as 'free' from a climate perspective even as it reduces land for all these other purposes."

48   sustainable fuels producer called LanzaJet: LanzaJet is one of around seventy industry players— airlines, airports, OEMs—that has spent years lobbying for a federal sustainable aviation fuel blenders credit—over and above the state-level credits that are already given to SAF users in California and elsewhere. They got their wish in July 2021, when the Biden administration included an undertaking to pay airlines $1.25 for each gallon of fuel that meets a 50 percent lifetime emissions reduction, rising to $1.75 a gallon for a 100 percent reduction, in its Build Back Better bill. "What's critical about this measure," Menotti says, "is that it is long term, and incentive-based approaches absolutely need to be long-term." When we met it was also—to Menotti's frustration— stalled in Congress, along with the bill's other clean energy incentives.

50   could lead to a reduction of emissions to 44.8 grams: Jim Spaeth, "Sustainable Aviation Fuels from Low-Carbon Ethanol Production," Office of Energy Efficiency & Renewable Energy, October 20, 2021, https://www .energy.gov/eere/bioenergy/articles /sustainable-aviation-fuels-low -carbon-ethanol-production.

CHAPTER TWO

53   enough energy to drive a car 130 kilometers or to heat a home for two days: Marco Alvera, *The Hydrogen Revolution: a Blueprint for the Future of Clean Energy* (Hodder Studio, 2021), p. 69.

54   hydrogen has been used for processing oil in refineries: Alvera, 80.

55   the cost of renewable hydrogen will drop by 30 percent: Mark Taylor and Dana Perkins, "Airbus on Ambitions for Zero Emissions Flight," *Bloomberg Switched On* podcast, November 18, 2020, https://podcasts.apple.com /gb/podcast/airbus-on-ambitions -for-zero-emissions-flight/id 1469286286?i=1000499196178.

56   there are ways of minimizing the effects of contrails: When you use a conventional gas turbine, soot particles in the exhaust behave as nucleation points for water vapor. With hydrogen, if any fuel impurities can be eliminated, this nucleation could be reduced significantly, although contrails might last longer because hydrogen gives off more water vapor than kerosene. Water vapor is much less harmful if it is given off in the lowest layer of the atmosphere, the troposphere, which might lead to constraints on the altitude at which planes can fly. Unlike hydrogen combustion aircraft, hydrogen fuel cell aircraft could be designed to store some of the water that is produced and release it in conditions that are less conducive to contrail/AIC formation (Roland Berger, 2020).

57 **"ultimate high performance hydrogen aircraft":** Mark Taylor and Dana Perkins, "Airbus on Ambitions for Zero Emissions Flight."

57 **Not forgetting, of course, the predictable need for taxpayer money:** Mark Taylor and Dana Perkins, "Airbus on Ambitions for Zero Emissions Flight."

58 **the repurposed A380 that is now the company's "hydrogen propulsion flight laboratory":** It's nice that the A380 is good for something.

61 **the same regional market that Airbus hopes to conquer:** An aircraft is considered regional if it has a capacity of around 100 seats and a range of up to 2,000 miles.

64 **Pink Floyd sent a massive inflatable pig:** Continuing eastward, causing consternation among pilots on their approach to the capital, the adventuresome porker drifted some way over the English Channel before finally coming to rest on a dairy farm in Kent.

65 **appeared on French television to vaunt the craft's qualities:** Sébastien Bougon (Flying Whales), "Flying Whales lève 122 millions

d'euros," YouTube video, June 30, 2022, https://www.youtube.com /watch?v=tLSCVI3SUCI.

66 **his company's Airlander airship will be eventually propelled by a hydrogen fuel cell:** Angela Hatwell, "Airlander Crash Lands After Second Flight," YouTube video, August 26, 2016, https://www.youtube.com/watch ?v=pvRTC5ISYgQ.

66 **The Airlander, he predicts:** We Are FINN, "HAV's Airlander Project 'Moving Really Fast' Towards Production and First Flight," YouTube video, June 30, 2022, https://www.youtube.com /watch?v=oNe3VCb2Rxw.

67 **doubts over their primary fuel don't only stem from its scarcity:** Ian Taylor, "The New Age of the Airship: Could Blimps Be the Future of Air Travel?" *BBC Science Focus*, June 9, 2021, https://www .sciencefocus.com/future -technology/the-new-age-of-the -airship-could-blimps-be-the -future-of-air-travel/.

CHAPTER THREE

70 **and they sued Archer for stealing its aircraft design:** Wisk alleged that a former employee, Jing Xue, improperly downloaded almost 5,000 data files onto a personal device that he subsequently gave to Archer after joining the company. The US Attorney's Office did not bring

charges against Xue, but Wisk did not retract its broader allegations of trade secret misappropriation and patent infringement, which are expected to come to trial in 2023. When I spoke to Goldstein, he dismissed the case as an irrelevance that would end with Archer's complete exoneration, but the awkward facts are that (a) Archer had poached more than a dozen former Wisk employees and (b) that in the one area of aviation where aircraft don't all look exactly the same, there's no denying the uncanny similarity between Archer's and Wisk's designs.

70 **according to a delightful company statement:** There are three broad categories of eVTOL aircraft: (1) multirotor, similar to helicopters but with multiple rotors distributed over an aircraft, generally without a fixed wing; (2) lift plus cruise, where one set of rotors are used for take-off and landing (vertical flight) and another set are used for cruising, generally with a fixed wing; and (3) vectored thrust, generally fixed-wing aircraft where the thrust-providing system of the aircraft is used in both vertical and forward flight by maneuvering the direction of thrust. Vectored thrust can be further categorized into (3a) tilt rotor, where rotors used in vertical flight tilt via rotating shafts to be used in forward flight; (3b) tilt wing, where the tilting action is performed by wings onto

which the rotors are attached; and (3c) tilt duct, similar to tilt rotor but with the thrust generated by propellers that are housed within cylindrical ducts, sometimes called ducted fans (Shashank Sripad and Venkatasubramanian Viswanathan, "The Promise of Energy-Efficient Battery-Powered Urban Aircraft," *Proceedings of the National Academy of Sciences*, Vol. 118, No. 45, November 9, 2021).

70 **the tantalizing promise of full automation:** The ethical, cultural, and technical aspects of automation will take years to sort out; for the foreseeable future eVTOLs will be piloted by someone on board, not by someone on the ground, far less by no one at all. And older people in particular will need time to get used to the idea that they don't have a shoulder to tap on when they want to say, "Excuse me, is the wing meant to be on fire?"

71 **"an affordable form of daily transportation for the masses":** Jeff Holden, Nikhil Goel, and Mark Moore, "Fast-Forwarding to a Future of On-Demand Urban Air Transportation," *Uber Elevate*, October 27, 2016.

74 **electric motors are between two and three times more efficient than combustion engines:** Sripad and Vaswanathan.

75 **all-electric aircraft would have a lifecycle carbon intensity 20 percent *higher*:** Schafer et al.

75   In India, taking an electric train can be more damaging to the climate than flying: Interview with Alan Epstein, July 11, 2022.

75   if you're running a plane using electricity from the Brazilian grid: Sofia Pinheiro Melo et al., "Life Cycle Engineering of Future Aircraft Systems: The Case of eVTOL Vehicles," *Procedia CIRP*, Vol. 90, 2020.

79   could switch to battery power during especially carbon-intensive phases of flight: Schafer et al.

80   will carry on getting lighter and more powerful: Lithium-ion batteries work by sending lithium ions from the negative electrode, the anode, through an electrolyte to the positive electrode, the cathode. The electrons stripped away at the anode travel toward the cathode along a circuit, creating a current than can power a motor. At the cathode, ions and electrons come together, the battery is plugged into a charger, and the process happens in reverse. Lithium-air batteries, on the other hand, use oxygen drawn from the atmosphere, generating more energy per unit of weight than the heavy metal oxides used in the lithium-ion model. "Solid-state" batteries are not only smaller and lighter still but also dispense with the flammable liquid or gel electrolyte that is found in the lithium-ion variety, reducing the risk of disagreeable conflagrations several miles up in the sky.

80   robots controlled by Venkat Viswanathan: David Stringer and Akshat Rathi, "Next-Generation Battery Pioneer Sees Breakthroughs Coming," Bloomberg, August 9, 2021, https://www.bloomberg.com /news/articles/2021-08-09/battery -pioneer-viswanathan-describes -the-breakthroughs-he-sees -coming.

80   capable of powering a single-aisle A320: Sripad and Vaswanathan.

CHAPTER FOUR

82   "100 million people in Asia will fly for the first time": Lizzy Gurdus, "Boeing CEO: Over 80% of the World Has Never Taken a Flight. We're Leveraging That for Growth," CNBC, December 7, 2017, https://www.cnbc.com/2017/12 /07/boeing-ceo-80-percent-of -people-never-flown-for-us-that -means-growth.html.

82   a vast and growing number of its citizens can afford plane tickets: China wants to become a big aircraft manufacturer and the government has funded the Commercial Aircraft Corporation of China, or COMAC, to achieve this goal. But production of COMAC's planes has been slow and they are inferior not only to Airbus and Boeing but also to Russian

110   planes like the Ilyushin, Sukhoi, and Tupolev. In development since 2008, COMAC's narrow-bodied airliner, the C19, which is intended to compete with the A320 and Boeing's 737, had yet to achieve certification in the summer of 2022. Nor is the C19 very Chinese. As Scott Kennedy, chair of Chinese business and economics at the Center for Strategic and International Studies, points out, "Everything that keeps it in the air—all the important parts—are from the United States and Europe." At its peak in 2018, the US-China aerospace trade balance stood at 17 to 1, with Europe enjoying similar success. China's reliance on foreign expertise—in June 2022 Airbus announced Chinese orders worth more than $37 billion—looks likely to continue.

83   **number of flights taking off from Chinese airports:** Jinglei Yu, et al., "China's Aircraft-Related CO2 Emissions: Decomposition Analysis, Decoupling Status, and Future Trends," *Energy Policy*, 2020.

83   **China would overtake the US as the world's biggest aviation market:** "China Becomes the Largest Aviation Market in the World," CAPA Centre for Aviation, April 16, 2020, https:// centreforaviation.com/analysis /reports/china-becomes-the -largest-aviation-market-in-the -world-521779.

83   **If UK aviation follows the "balanced net-zero pathway":** The UK's modeling is conservative when it comes to new technologies, assuming that the uptake of electric hybrid aircraft will be relatively modest (around 9 percent of aircraft kilometers by 2050, consuming 6 to 7 percent of jet fuel), and that full electric planes will not be commercialized by the midpoint of the century. Nor, in the committee's view, can hydrogen turbine or hydrogen fuel cell planes be expected to take the strain; the committee noted the "significant" time it takes "to design, build, test, scale-up, certify, and manufacture new aircraft propulsion systems (and the new aircraft bodies to accommodate them and their energy stores on-board)," https:// www.theccc.org.uk/wp-content /uploads/2020/12/Sector-summary -Aviation.pdf.

84   **three scenarios are possible for the country's aviation emissions:** Yu et al.

85   **A resolution passed by the same body in 2016:** "Resolution A39-2: Consolidated Statement of Continuing ICAO Policies and Practices Related to Environmental Protection— Climate Change," https://www .icao.int/environmental-protection /documents/resolution_a39_2.pdf.

86   **The majority of China's offsets:** Barbara Finamore, "China a Key Player in Aviation Emissions

Agreement," September 13, 2016, NRDC Expert Blog, https://www.nrdc.org/experts/barbara-finamore/china-key-player-aviation-emissions-agreement.

86 **Even the International Maritime Organization:** "Setting a Long-Term Climate Change Goal for International Aviation," International Civil Aviation Organization, working paper, June 8, 2019, https://www.icao.int/Meetings/a40/Documents/WP/wp_277_en.pdf.

89 **"a two-point strategy . . . to avoid regulation":** "The Aviation Industry and European Climate Policy," an InfluenceMap report, June 2021, https://influencemap.org/report/Aviation-Industry-Lobbying-European-Climate-Policy-131378131d9503b4d32b365e54756351.

CONCLUSION

92–93 **the most durable and limitless method we have to fix carbon:** A lot more durable than planting trees that can die with the first winter or burn with the first spark.

93 **a particularly pernicious form of greenwashing:** In 2020, United Airlines announced that, as part of its plans to go "100 percent green" by 2050, it would mitigate its emissions by investing in a DAC facility producing CO_2 to help Occidental Petroleum extract oil from a field in Texas. That drilling for oil and then refining and burning its products is very far from being "100 percent green" seems to have eluded the company.

93 **the IPCC acknowledged this reality in 2022:** "Climate Change 2022: Mitigation of Climate Change," Intergovernmental Panel on Climate Change, 2022, https://www.ipcc.ch/report/ar6/wg3/downloads/report/IPCC_AR6_WGIII_SPM.pdf.

94 **it's also part of the most pessimistic scenario:** At the time of the publication of the IEA's report, "Direct Air Capture: A Key Technology for Net Zero," in 2022, just eighteen small DAC plants were in operation around the world, in Canada, Europe, and the United States. But Climeworks's Christoph Gebald is adamant that DAC can grow into a trillion-dollar industry in the next three to four decades. "We want to work at the maximum speed that humanity has shown to date in terms of technology scale-up," he told me, "and this is . . . 1,000x. Both wind and solar did that in the last twenty years." All this for a product that is a form of waste disposal.

Columbia Global Reports is a publishing imprint from Columbia University that commissions authors to produce works of original thinking and on-site reporting from all over the world, on a wide range of topics. Our books are short—novella-length, and readable in a few hours—but ambitious. They offer new ways of looking at and understanding the major issues of our time. Most readers are curious and busy. Our books are for them.

Subscribe to our newsletter, and learn more about Columbia Global Reports at globalreports.columbia.edu.